Texts and Readings in Mathematics

Volume 75

The *Texts and Readings in Mathematics* series publishes high-quality textbooks, research-level monographs, lecture notes and contributed volumes. Undergraduate and graduate students of mathematics, research scholars, and teachers would find this book series useful. The volumes are carefully written as teaching aids and highlight characteristic features of the theory. The books in this series are co-published with Hindustan Book Agency, New Delhi, India.

More information about this series at http://www.springer.com/series/15141

Arup Bose · Snigdhansu Chatterjee

U-Statistics, M_m-Estimators and Resampling

 HINDUSTAN
BOOK AGENCY

 Springer

Arup Bose
Statistics and Mathematics Unit
Indian Statistical Institute
Kolkata, West Bengal, India

Snigdhansu Chatterjee
School of Statistics
University of Minnesota
Minneapolis, MN, USA

ISSN 2366-8717   ISSN 2366-8725   (electronic)
Texts and Readings in Mathematics
ISBN 978-981-13-4756-6   ISBN 978-981-13-2248-8   (eBook)
https://doi.org/10.1007/978-981-13-2248-8

This Springer imprint is published by the registered company Springer Nature Singapore Pte Ltd.
The registered company address is: 152 Beach Road, #21-01/04 Gateway East, Singapore 189721, Singapore

Contents

Contents

To Chhoti and Mita

A. B.

To Baishali

S. C.

Preface

This small book covers three important topics that we believe every statistics student should be familiar with: U-statistics, M_m-estimates and Resampling. The final chapter is a quick and short introduction to the statistical software R, primarily geared towards implementing resampling. We hasten to add that the book is introductory. However, adequate references are provided for the reader to explore further.

Any U-statistic (with finite variance) is the non-parametric minimum variance unbiased estimator of its expectation. Many common statistics and estimators are either U-statistics or are approximately so. The systematic study of U-statistics began with Hoeffding (1948) and comprehensive treatment of U-statistics are available at many places including Lee (1990) and Korolyuk and Borovskich (1993).

In Chapter 1 we cover the very basics of U-statistics. We begin with its definition and examples. The exact finite sample distribution and other properties of U-statistics can seldom be calculated. We cover some asymptotic properties of U-statistics such as the central limit theorem, weak and strong law of large numbers, law of iterated logarithm, a deviation result and, a distribution limit theorem for a first order degenerate U-statistics. As direct application of these, we establish the asymptotic normality of many common estimators and the sum of weighted chi-square limit for the Cramér-von Mises statistic. Other applications are provided in Chapter 2. In particular the idea of linearization or the so called weak representation of a U-statistic carries over to the next chapters.

Chapter 2 is on M-estimators and their general versions M_m-estimators, introduced by Huber (1964) out of robustness considerations. Asymptotic properties of these estimates have been treated under different sets of conditions in several books and innumerable research articles. Establishing the most general results for these estimators requires sophisticated treatment using techniques from the theory of empirical processes. We strive for a simple approach.

We impose a few very simple conditions on the model. Primary among these is a convexity condition which is still general enough to be applicable widely. Under these conditions, a huge class of M_m-estimators are approximate U-statistics. Hence the theory developed in Chapter 1 can be profitably used to derive asymptotic properties, such as the central limit theorem for M_m-estimators by linearizing them. We present several examples to show how the general results can be applied to specific estimators. In particular, several multivariate estimates of location are discussed in details.

The linearization in Chapters 1 and 2 was achieved by expending considerable technical effort. There still remain two noteworthy issues. First, such a linearization may not be easily available for many estimates. Second, even if an asymptotic normality result is established, it may not be easy to find or estimate the asymptotic variance.

Since the inception of the bootstrap method in the early eighties, an alternative to asymptotic distributional results is Resampling, This is now a necessary item in the everyday toolkit of a statistician. It attempts to replace analytic derivations with the force of computations. Again, there are several excellent monographs and books on this topic, in addition to the surfeit of articles on both theory and applications of Resampling.

In Chapter 3, we introduce the main ideas of resampling in an easy way by using three benchmark examples; sample mean, sample median and ordinary least squares estimates of regression parameters. In particular, we also explain when and how resampling can produce "better" estimates than those from traditional asymptotic normal approximations. We also present a short overview of the most common methods of resampling.

Chapter 4 focuses on resampling estimates for the sampling distribution and asymptotic variance of U-statistics and M_m-estimators. In particular, we discuss the traditional Efron's (multinomial) bootstrap and its drawbacks in the context of U-statistics. We discuss how the generalized bootstrap arises naturally in this context. We establish a bootstrap linearization result for U-statistics. The generalized bootstrap with additive and multiplicative weights are given special attention, the first due to the computational savings obtained and the second due to its connection with Efron's bootstrap. Finally, we also establish a weighted U-statistics result for the generalized bootstrap M_m-

estimates. These linearization results show the distributional consistency of the generalized bootstrap for U-statistics and M_m-estimates.

Chapter 5 brings in an additional flavor: that of computation and practical usage of the various estimation and inferential techniques discussed. We do not assume *any* knowledge about statistical softwares or computational practices on the part of the reader, hence this chapter is suitable for a beginner. We describe various aspects of the free and versatile statistical software R, followed by a very brief illustration on how to practically use some of the techniques presented in this book. We created a free and publicly available R package called UStatBookABSC (Chatterjee (2016)) to accompany this book, which also contains a dataset that we use for illustrating various R commands and concepts.

The works of SC and AB have been partially supported respectively by the United States National Science Foundation (NSF) under the grants # DMS-1622483 and # DMS-1737918 and by the J.C. Bose Fellowship, Govt. of India. We thank Monika Bhattacharjee, Gouranga Chatterjee and Deepayan Sarkar for proof reading parts of the book, and for providing valuable comments. We are thankful to Lindsey Dietz for obtaining and curating the data on precipitation and temperature that we have used. We thank Rajendra Bhatia for encouraging us to write in the TRIM series.

Arup Bose, Kolkata and Snigdhansu Chatterjee, Minneapolis

April 15, 2018

About the Authors

Arup Bose is Professor at the Statistics and Mathematics Unit, Indian Statistical Institute, Kolkata, India. He is a Fellow of the Institute of Mathematical Statistics and of all the three national science academies of India. He has significant research contributions in the areas of statistics, probability, economics and econometrics. He is a recipient of the Shanti Swarup Bhatnagar Prize and the C R Rao National Award in Statistics. His current research interests are in large dimensional random matrices, free probability, high dimensional data, and resampling. He has authored three books: *Patterned Random Matrices,* and *Large Covariance and Autocovariance Matrices* (with Monika Bhattacharjee) and *Random Circulant Matrices* (with Koushik Saha), published by Chapman & Hall.

Snigdhansu Chatterjee is Professor at the School of Statistics, University of Minnesota, USA. He is also the Director of the Institute for Research in Statistics and its Applications. His research interests are in resampling methods, high-dimensional and big data statistical methods, small area methods, and application of statistics in climate science, neuroscience and social sciences. He has written over 45 research articles.

Chapter 1

Introduction to U-statistics

U statistics are a large and important class of statistics. Indeed, any U-statistic (with finite variance) is the non-parametric minimum variance estimator of its expectation θ. Many common statistics and estimators are either U-statistics or approximately so. The systematic study of U-statistics began with Hoeffding (1948). Many of the basic properties of U-statistics are due to him. Full length treatment of U-statistics are provided by Lee (1990) and Korolyuk and Borovskich (1993).

Our goal in this chapter is to cover the very basics of U-statistics in a concise and more or less self-contained way. We will start with the definition of a U-statistic and provide several examples. We will then cover some asymptotic properties of U-statistics such as the central limit theorem, weak and strong law of large numbers, law of iterated logarithm, and a distribution limit theorem for a first order degenerate U-statistic. As applications of these results, we will establish the asymptotic distribution of many common statistics, including the Cramér-von Mises statistic. In the next chapter we shall see that there is a general class of statistics, called M_m estimators, which are also approximate U-statistics. Thus the results of this chapter will be applicable to an even wider class of statistics.

1.1 Definition and examples

Let $Y_1, Y_2, \ldots, Y_n$ be $\mathcal{Y}$-valued random variables. It will be clear from the context what $\mathcal{Y}$ is. We shall assume throughout that they are independent

© Springer Nature Singapore Pte Ltd. 2018 and Hindustan Book Agency 2018
A. Bose and S. Chatterjee, *U-Statistics, Mm-Estimators and Resampling*, Texts and Readings in Mathematics 75, https://doi.org/10.1007/978-981-13-2248-8_1

and identically distributed (i.i.d. hereafter). Suppose $h(x_1, \ldots, x_m)$ is a real valued function on $\mathcal{Y}^m$ which is *symmetric in its arguments*.

Definition 1.1: The U-statistic of *order* or *degree* m, with *kernel* h is defined as:

$$U_n(h) = \binom{n}{m}^{-1} \sum_{1 \leq i_1 < \cdots < i_m \leq n} h(Y_{i_1}, \ldots, Y_{i_m}). \qquad (1.1)$$

We shall often write U_n for $U_n(h)$ when the function h is clear from the context. Appropriate extension is available when h is vector valued.

Note that a U-statistic of degree m is also a U-statistic of degree $(m+1)$. As a consequence, the sum of two U-statistics is again a U-statistic. In this book, we consider the order m to be the smallest integer for which the above definition holds.

Example 1.1(*Sample mean*): Let $Y_1, Y_2, \ldots, Y_n$ be observations, then with $m = 1$ and $h(x) = x$, we obtain

$$U_n = n^{-1} \sum_{i=1}^{n} Y_i = \overline{Y}. \qquad (1.2)$$

□

Example 1.2(*Sample variance*): Let $Y_1, Y_2, \ldots, Y_n$ be the observations, then with $m = 2$ and $h(x_1, x_2) = \frac{(x_1 - x_2)^2}{2}$, we get

$$U_n = \binom{n}{2}^{-1} \sum_{1 \leq i_1 < i_2 \leq n} (Y_{i_1} - Y_{i_2})^2 / 2. \qquad (1.3)$$

It is easily seen that U_n is the *sample variance*,

$$U_n = (n-1)^{-1} \sum_{i=1}^{n} (Y_i - \overline{Y})^2 = s_n^2. \qquad (1.4)$$

□

Example 1.3(*Sample covariance*): Suppose (X_i, Y_i), $1 \leq i \leq n$ are the observations, and consider $m = 2$ and

$$h\big((x_1, y_1), (x_2, y_2)\big) = \frac{1}{2}(x_1 - x_2)(y_1 - y_2).$$

Then U_n is the *sample covariance* of $\{(X_i, Y_i)\}$, given by

$$U_n = (n-1)^{-1} \sum_{i=1}^{n} (X_i - \overline{X})(Y_i - \overline{Y}). \tag{1.5}$$

$\square$

Example 1.4*(Kendall's tau)*: Define the sign function as

$$\text{sign}(x) = \begin{cases} -1 & \text{if } x < 0, \\ 0 & \text{if } x = 0, \\ 1 & \text{if } x > 0. \end{cases}$$

Suppose (X_i, Y_i), $1 \leq i \leq n$ are continuous bivariate observations. *Kendall's tau*, a measure of *discordance*, is defined by

$$t_n = \binom{n}{2}^{-1} \sum_{1 \leq i < j \leq n} \text{sign}((X_i - X_j)(Y_i - Y_j)). \tag{1.6}$$

This is a U-statistic with $h((x_1, x_2), (y_1, y_2)) = \text{sign}((x_1 - x_2)(y_1 - y_2))$. $\square$

Example 1.5*(Gini's mean difference)*: A measure of income inequality is the *Gini's mean difference* given by

$$U_n = \binom{n}{2}^{-1} \sum_{1 \leq i < j \leq n} |Y_i - Y_j|. \tag{1.7}$$

If the observations are i.i.d. $N(0, \sigma^2)$, then we have

$$\mathbb{E}(U_n) = \frac{2}{\pi^{1/2}} \sigma. \tag{1.8}$$

Thus U_n is a U-statistic with $h(x_1, x_2) = |x_1 - x_2|$ and is a measure of dispersion. $\square$

Example 1.6*(Wilcoxon's one sample rank statistic)*: Suppose Y_i, $1 \leq i \leq n$, are continuous observations. Let $R_i = \text{Rank} (|Y_i|)$, $1 \leq i \leq n$. *Wilcoxon's one sample rank statistic* is defined as

$$T^+ = \sum_{i=1}^{n} R_i \, \mathcal{I}_{\{Y_i > 0\}}. \tag{1.9}$$

T^+ is the sum of the ranks of all the positive observations. It can be written as a linear combination of two U-statistics with kernels of sizes 1 and 2. To do this, note that for $i \neq j$,

$$\mathcal{I}_{\{Y_i+Y_j>0\}} = \mathcal{I}_{\{Y_i>0\}}\mathcal{I}_{\{|Y_j|<Y_i\}} + \mathcal{I}_{\{Y_j>0\}}\mathcal{I}_{\{|Y_i|<Y_j\}}. \qquad (1.10)$$

Hence with probability one

$$\sum_{1\leq i\leq j\leq n} \mathcal{I}_{\{Y_j+Y_j>0\}}$$

$$= \sum_{1\leq i<j\leq n} \mathcal{I}_{\{Y_i>0\}}\mathcal{I}_{\{|Y_j|<Y_i\}} + \sum_{1\leq i<j\leq n} \mathcal{I}_{\{Y_j>0\}}\mathcal{I}_{\{|Y_i|<Y_j\}} + \sum_{i=1}^{n}\mathcal{I}_{\{Y_i>0\}}$$

$$= \sum_{i=1}^{n}\sum_{j=1}^{n} \mathcal{I}_{\{Y_i>0\}}\mathcal{I}_{\{|Y_j|\leq Y_i\}}$$

$$= \sum_{i=1}^{n}\mathcal{I}_{\{Y_i>0\}}R_i = T^+.$$

It is now easy to see that

$$T^+ = \binom{n}{2}U_n(f) + nU_n(g) \qquad (1.11)$$

where $U_n(f)$ and $U_n(g)$ are the U-statistics

$$U_n(f) = \binom{n}{2}^{-1}\sum_{1\leq i<j\leq n} f(Y_i, Y_j) \text{ and } U_n(g) = \frac{1}{n}\sum_{i=1}^{n} g(Y_i)$$

with kernels $f(x_1, x_2) = \mathcal{I}_{\{x_1+x_2>0\}}$ and $g(x_1) = \mathcal{I}_{\{x_1>0\}}$. $\qquad\square$

Example 1.7: A correlation coefficient different from the usual product moment correlation was introduced and studied in details by Bergsma (2006). We need some preparation to define it. First suppose that Z, Z_1 and Z_2 are i.i.d. real valued random variables with the distribution function F. Let

$$h_F(z_1, z_2) = -\frac{1}{2}\mathbb{E}\big(|z_1 - z_2| - |z_1 - Z_2| - |Z_1 - z_2| + |Z_1 - Z_2|\big). \qquad (1.12)$$

Note that if Z_3, Z_4 are i.i.d. F, then

$$\mathbb{E}h_F(Z_3, Z_4) = 0. \qquad (1.13)$$

Now let (X, Y) be a bivariate random variable with marginal distributions F_X and F_Y. Let (X_1, Y_1), and (X_2, Y_2) be i.i.d. copies of (X, Y). Then a "covariance" between X and Y is defined by

$$\kappa(X,Y) = \mathbb{E}h_{F_X}(X_1, X_2)h_{F_Y}(Y_1, Y_2)$$

and a "correlation coefficient" between them is defined by

$$\rho^*(X,Y) = \frac{\kappa(X,Y)}{\sqrt{\kappa(X,X)\kappa(Y,Y)}}.$$

It is easy to see that if X and Y are independent then $\kappa(X,Y) = \rho^*(X,Y) = 0$. It turns out that $0 \le \rho^* \le 1$ for any pair of random variables (X,Y) and $\rho^*(X,Y) = 0$ if and only if X and Y are independent. In addition, $\rho^*(X,Y) = 1$ if and only if X and Y are linearly related. See Bergsma (2006) for details.

Now suppose we have n observations $(X_i, Y_i), 1 \le i \le n$ from a bivariate distribution $F_{X,Y}$. Based on these observations, we wish to test the hypothesis

$$H_0 : F_{X,Y}(x,y) = F_X(x)F_Y(y) \text{ for all } x, y. \tag{1.14}$$

Since $\kappa(X,Y) = 0$ if and only if H_0 is true, we can base our test on a suitable estimate of $\kappa(X,Y)$. The most reasonable estimate is given using a U-statistic, which we describe below. Define the following quantities:

$$A_{1k} = \frac{1}{n}\sum_{i=1}^{n}|X_k - X_i|, \quad A_{2k} = \frac{1}{n}\sum_{i=1}^{n}|Y_k - Y_i|, \text{ and}$$

$$B_1 = \frac{1}{n^2}\sum_{i=1}^{n}\sum_{j=1}^{n}|X_i - X_j|, \quad B_2 = \frac{1}{n^2}\sum_{i=1}^{n}\sum_{j=1}^{n}|Y_i - Y_j|.$$

For $k, l = 1, \dots, n$, let

$$h_{\hat{F}_X}(X_k, X_l) = -\frac{1}{2}\left(|X_k - X_l| - \frac{n}{n-1}A_{1k} - \frac{n}{n-1}A_{1l} + \frac{n}{n-1}B_1\right),$$

$$h_{\hat{F}_Y}(Y_k, Y_l) = -\frac{1}{2}\left(|Y_k - Y_l| - \frac{n}{n-1}A_{2k} - \frac{n}{n-1}A_{2l} + \frac{n}{n-1}B_2\right).$$

Then an unbiased estimator of $\kappa(X,Y)$ is given by

$$\kappa_n = \binom{n}{2}^{-1}\sum_{1 \le i < j \le n}h_{\hat{F}_X}(X_i, X_j)h_{\hat{F}_Y}(Y_i, Y_j).$$

Note that κ_n may be negative even though κ is never so. Clearly κ is a U-statistic of degree 2. □

1.2   Some finite sample properties

1.2.1   Variance

Assume that $\mathbb{V}\big(h(Y_1,\ldots,Y_m)\big) < \infty$ where $\mathbb{V}$ denotes variance. To compute $\mathbb{V}(U_n)$, we need to compute

$$\mathbb{COV}\big(h(Y_{j_1},\ldots,Y_{j_m}),h(Y_{i_1},\ldots,Y_{i_m})\big)$$

where $\mathbb{COV}$ denotes covariance.

Let c denote the number of common indices between $\{i_1,\ldots,i_m\}$ and $\{j_1,\ldots,j_m\}$. The total number of such covariance terms equals

$$\binom{n}{m}\binom{m}{c}\binom{n-m}{m-c}.$$

This can be seen as follows. First, we can choose m indices $\{i_1,\ldots,i_m\}$ from $\{1,\ldots,n\}$ in $\binom{n}{m}$ ways. Then choose c of those which are to be common with $\{j_1,\ldots,j_m\}$ in $\binom{m}{c}$ ways. Now choose the rest of the $(m-c)$ indices of $\{j_1,\ldots,j_m\}$ from the $(n-m)$ remaining indices in $\binom{n-m}{m-c}$ ways.

For any c, $1 \le c \le m$ define

$$\delta_c = \mathbb{COV}\big[h(Y_{i_1},\ldots,Y_{i_m}),h(Y_{j_1},\ldots,Y_{j_m})\big],$$
$$h_c(x_1,\ldots,x_c) = \mathbb{E}\big[h(Y_1,\ldots,Y_c,Y_{c+1},\ldots,Y_m)|Y_1 = x_1,\ldots,Y_c = x_c\big].$$

Suppose $\{\tilde{Y}_1,\tilde{Y}_2,\ldots\}$ is an i.i.d. copy of $\{Y_1,Y_2,\ldots\}$ (without loss of generality defined on the same probability space). Then clearly,

$$\delta_c = \mathbb{COV}\big[h(Y_1,\ldots,Y_c,Y_{c+1},\ldots,Y_m),h(Y_1,\ldots,Y_c,\tilde{Y}_{c+1},\ldots,\tilde{Y}_m)\big]$$
$$= \mathbb{E}\Big[h(Y_1,\ldots,Y_c,Y_{c+1},\ldots,Y_m)h(Y_1,\ldots,Y_c,\tilde{Y}_{c+1},\ldots,\tilde{Y}_m)\Big]$$
$$- \Big(\mathbb{E}\big[h(Y_1,\ldots,Y_m)\big]\Big)^2$$
$$= \mathbb{E}\Big[h_c(Y_1,\ldots,Y_c)\Big]^2 - \Big(\mathbb{E}\big[h_c(Y_1,\ldots,Y_c)\big]\Big)^2$$
$$= \mathbb{V}\Big[h_c(Y_1,\ldots,Y_c)\Big] \ge 0.$$

Hence

$$\mathbb{V}(U_n)$$

$$= \binom{n}{m}^{-2} \mathrm{COV}\Big(\sum_{1 \le i_1 < \cdots < i_m \le n} h(Y_{i_1}, \ldots, Y_{i_m}), \sum_{1 \le j_1 < \cdots < j_m \le n} h(Y_{j_1}, \ldots, Y_{j_m}) \Big)$$

$$\tag{1.15}$$

$$= \binom{n}{m}^{-2} \sum_{c=1}^{m} \binom{n}{m} \binom{m}{c} \binom{n-m}{m-c} \delta_c$$

$$= \binom{n}{m}^{-1} \sum_{c=1}^{m} \binom{m}{c} \binom{n-m}{m-c} \delta_c. \tag{1.16}$$

As a consequence, as $n \to \infty$,

$$\mathbb{V}(U_n) = \frac{m^2 \delta_1}{n} + O(n^{-2}) \quad \text{and} \tag{1.17}$$

$$\mathbb{V}\big(n^{1/2}(U_n - \theta)\big) \to m^2 \delta_1. \tag{1.18}$$

Example 1.8: Let $h(x_1, x_2) = \frac{(x_1 - x_2)^2}{2}$. Then

$$U_n = s_n^2 = (n-1)^{-1} \sum_{i=1}^{n} (Y_i - \overline{Y}_n)^2.$$

It can be verified that, if $\sigma^2 = \mathbb{V}(Y_1)$ and $\mu_4 = \mathbb{E}(Y - \mathbb{E}(Y))^4$, then

$$\delta_1 = \frac{\mu_4 - \sigma^4}{4}, \tag{1.19}$$

$$\delta_2 = \frac{\mu_4 + \sigma^4}{2}, \quad \text{and}$$

$$\mathbb{V}(U_n) = \frac{4(n-2)}{n(n-1)} \delta_1 + \frac{2}{n(n-1)} \delta_2. \tag{1.20}$$

$$\square$$

1.2.2   First projection

The *first projection* of a U-statistic U_n, denoted by $h_1(\cdot)$, is the *conditional expectation* of h given *one* of the coordinates:

$$h_1(x_1) = \mathbb{E}\, h(x_1, Y_2, \ldots, Y_m) \tag{1.21}$$

Define the centered version of the first projection:

$$\tilde{h}_1(x_1) = h_1(x_1) - \theta$$

so that

$$\mathbb{E}\,\tilde{h}_1(Y_1) = \mathbb{E}\big[h(Y_1, \ldots, Y_m)\big] - \theta = 0.$$

Let

$$R_n = U_n - \theta - \frac{m}{n}\sum_{i=1}^{n}\tilde{h}_1(Y_i). \tag{1.22}$$

By explicit calculations it can be easily seen that the decomposition (1.22) is an *orthogonal decomposition* in the following sense:

$$\mathbb{COV}[\tilde{h}_1(Y_i), R_n] = 0 \quad \forall\, i = 1, \ldots, n. \tag{1.23}$$

Notice that in (1.18) the leading term in $\mathbb{V}(n^{1/2}U_n)$ equals $m^2\delta_1$ where

$$\delta_1 = \mathbb{COV}\big[h(Y_1, Y_2, \ldots, Y_m), h(Y_1, \tilde{Y}_2, \ldots, \tilde{Y}_m)\big] \tag{1.24}$$

$$= \mathbb{V}\big(h_1(Y_1)\big). \tag{1.25}$$

1.3 Law of large numbers and asymptotic normality

The notation $\xrightarrow{\mathcal{D}}$, $\xrightarrow{\mathbb{P}}$ and $\xrightarrow{a.s.}$ shall be used throughout the book to denote convergence in distribution, convergence in probability and convergence almost surely, respectively.

From the relations (1.22), (1.23) and (1.18), we get

$$\mathbb{V}(U_n) = \frac{m^2}{n}\delta_1 + \mathbb{V}(R_n)$$

$$= \frac{m^2}{n}\delta_1 + O(n^{-2}). \tag{1.26}$$

This shows that

$$\mathbb{V}(n^{1/2}R_n) \to 0 \quad \text{and hence} \quad n^{1/2}R_n \xrightarrow{\mathbb{P}} 0. \tag{1.27}$$

Now we obtain two important results on U-statistics. Theorem 1.1(b) is the *Central Limit Theorem* for U-statistics (UCLT). It follows by appealing to the decomposition given in part (a) and the central limit theorem (CLT) for the sample mean of i.i.d. observations. Theorem 1.1(a) is a *weak representation* or *linearization* of U-statistics and follows from (1.22) and (1.27). This is useful when we have to deal with several U-statistics simultaneously. Then it is easy to see that the *multivariate* version of Theorem 1.1(b) holds.

In Theorem 1.2, using Theorem 1.1(a), we obtain the *weak law of large numbers* (WLLN) for U-Statistics.

Theorem 1.1 (Hoeffding (1948)). *(UCLT.)* If $\mathbb{V}[h(Y_1, \ldots, Y_m)] < \infty$, then

(a)

$$U_n - \theta = \frac{m}{n} \sum_{i=1}^{n} \tilde{h}_1(Y_i) + R_n \quad \text{where} \quad n^{1/2} R_n \xrightarrow{\mathbb{P}} 0. \quad (1.28)$$

(b)

$$n^{1/2}(U_n - \theta) \xrightarrow{\mathcal{D}} N(0, m^2 \delta_1) \quad \text{where} \quad \delta_1 = \mathbb{V}(\tilde{h}_1(Y_1)).$$

Theorem 1.2. *(Weak law of large numbers (WLLN) for U-statistics.)* If $\mathbb{V}[h(Y_1, \ldots, Y_m)] < \infty$, then

$$U_n - \theta \xrightarrow{\mathbb{P}} 0 \text{ as } n \to \infty.$$

Rate results in the weak law, when stronger moment conditions are assumed, are given in Section 1.4. A much stronger result than the weak law is actually true for U-statistics and we state it below:

Theorem 1.3 (Hoeffding (1961)). *(Strong law of large numbers (SLLN) for U-statistics.)* If $\mathbb{E}\big[|h(Y_1, \ldots, Y_m)|\big] < \infty$, then

$$U_n - \theta \xrightarrow{a.s.} 0 \text{ as } n \to \infty.$$

This result can be proved by using SLLN for either reverse martingales Berk (1966) or forward martingales Hoeffding (1961). See Lee (1990) for a detailed proof.

The next result we present is the *Law of iterated logarithms* (LIL) for U-statistics. See Lee (1990) for a proof of this result. This result will be used in the proof of Theorem 2.5 in Chapter 2.

Theorem 1.4 (Dehling et al. (1986)). *(Law of iterated logarithm (LIL) for U-statistics.) Suppose U_n is a U-statistic with kernel h such that $\delta_1 > 0$ and $\mathbb{E}\big[|h(Y_1, \ldots, Y_m)|^2\big] < \infty$. Then as $n \to \infty$,*

$$\limsup_{n} \frac{n(U_n - \theta)}{\sqrt{2m^2 \delta_1 \log \log n}} = 1 \; almost \; surely.$$

Example 1.9: Consider the U-statistic s_n^2. In Example 1.8, we have calculated

$$\delta_1 = \frac{\mu_4 - \sigma^4}{4},$$

where $\mu_4 = \mathbb{E}\big(Y - (\mathbb{E}Y)\big)^4$. Thus if $\mu_4 < \infty$,

$$n^{1/2}(s_n^2 - \sigma^2) \xrightarrow{\mathcal{D}} N(0, \, \mu_4 - \sigma^4).$$

Note that when $\mu_4 = \sigma^4$, the limit is degenerate at 0. □

Example 1.10: Suppose $h(x_1, x_2) = x_1 x_2$, $\sigma^2 = \mathbb{V}(Y_1) < \infty$ and $\mu = \mathbb{E}(Y_1)$. Then

$$\begin{aligned}
\delta_1 &= \mathbb{COV}(Y_1 Y_2, Y_1 \tilde{Y}_2) \\
&= \mathbb{E}Y_1^2 Y_2 \tilde{Y}_2 - \big(\mathbb{E}Y_1 Y_2\big)\big(\mathbb{E}Y_1 \tilde{Y}_2\big) \\
&= \mu^2 \mathbb{E}Y_1^2 - \mu^4 = \mu^2 \mathbb{V}(Y_1) = \mu^2 \sigma^2.
\end{aligned}$$

Hence, since $m = 2, \delta_1 = \mu^2 \sigma^2$, we have

$$n^{1/2}\left(\binom{n}{2}^{-1} \sum_{1 \le i < j \le n} Y_i Y_j - \mu^2\right) \xrightarrow{\mathcal{D}} N(0, 4\,\mu^2 \sigma^2).$$

Note that when $\mu = 0$, the limit distribution is *degenerate*. We shall deal with general degenerate U-statistics in the next section. □

Example 1.11: Consider Kendall's tau defined in Example 1.4. It is used to test the *null* hypothesis that X and Y are independent. We shall derive the asymptotic null distribution of this statistic. Let $\mathbb{F}, \mathbb{F}_1$ and $\mathbb{F}_2$ denote the distributions of (X, Y), X and Y, respectively. Assume that they are continuous. Then

$$h_1(x, y) \tag{1.29}$$
$$= \mathbb{E}\, h((x, y), (X_2, Y_2))$$
$$= \mathbb{P}\big[(x - X_2)(y - Y_2) > 0\big] - \mathbb{P}\big[(x - X_2)(y - Y_2) < 0\big]$$
$$= \mathbb{P}\big[(X_2 > x, Y_2 > y), \text{ or } (X_2 < x, Y_2 < y)\big]$$
$$- \mathbb{P}\big[(X_2 > x, Y_2 < y) \text{ or } (X_2 < x, Y_2 > y)\big]$$
$$= 1 - 2\mathbb{F}(x, \infty) - 2\mathbb{F}(\infty, y) + 4\,\mathbb{F}(x, y)$$
$$= \big(1 - 2\mathbb{F}_1(x)\big)\big(1 - 2\mathbb{F}_2(y)\big) + 4\big(\mathbb{F}(x, y) - \mathbb{F}_1(x)\mathbb{F}_2(y)\big). \tag{1.30}$$

Under the null hypothesis,

$$\mathbb{F}(x, y) = \mathbb{F}_1(x)\mathbb{F}_2(y) \text{ for all } x, y. \tag{1.31}$$

Hence in that case,

$$h_1(x, y) = \big(1 - 2\mathbb{F}_1(x)\big)\big(1 - 2\mathbb{F}_2(y)\big). \tag{1.32}$$

To compute δ_1, since $\mathbb{F}_1$ and $\mathbb{F}_2$ are continuous, under the null hypothesis,

$$U = 1 - 2\mathbb{F}_1(X_1) \quad \text{and} \quad V = 1 - 2\mathbb{F}_2(Y_1) \tag{1.33}$$

are independent $U(-1, 1)$ random variables. Hence

$$\delta_1 = \mathbb{V}[h_1(X, Y)] = \mathbb{V}(UV)$$
$$= \mathbb{E}U^2\mathbb{E}V^2 = \Big(\frac{1}{2}\int_{-1}^{1} u^2 du\Big)^2 = 1/9.$$

Since $m = 2$, we get $m^2\delta_1 = 4/9$. Moreover, under independence,

$$\theta = \mathbb{E}\Big[\, \text{sign}\,\big((X_1 - X_2)(Y_1 - Y_2)\big)\Big] = 0.$$

Hence *under independence,* $n^{1/2}U_n \xrightarrow{\mathcal{D}} N(0, 4/9)$. $\qquad\square$

Example 1.12: Wilcoxon's statistic, defined in Example 1.6 is used for testing the null hypothesis that the distribution $\mathbb{F}$ of Y_1 is continuous and symmetric about 0. Recall the expression for T^+ in (1.11). Note that $\mathbb{E}\, U_n(f) =$

$\mathbb{P}(Y_1 + Y_2 > 0) = \theta$ (say). Under the null hypothesis, $\theta = 1/2$. Further,

$$n^{-3/2}\sqrt{12}(T^+ - n^2/4) \qquad (1.34)$$

$$= n^{-3/2}\sqrt{12}\left(\binom{n}{2}U_n(f) + nU_n(g) - n^2/4\right)$$

$$= n^{-3/2}\sqrt{12}nU_n(g) + n^{1/2}\sqrt{3}(U_n(f) - 1/2) - n^{-1/2}\sqrt{3}U_n(f). \qquad (1.35)$$

Clearly, using the weak representation (1.28),

$$n^{-3/2}(nU_n(g)) \xrightarrow{\text{P}} 0. \qquad (1.36)$$

Further,

$$\mathbb{V}(\tilde{f}_1) = \mathbb{COV}\left[\mathcal{I}_{\{Y_1+Y_2>0\}}, \ \mathcal{I}_{\{Y_1+\tilde{Y}_2>0\}}\right]$$

$$= \mathbb{P}\left[Y_1 + Y_2 > 0, \ Y_1 + \tilde{Y}_2 > 0\right] - (1/2)^2.$$

Note that $\mathbb{F}$ *is continuous*. Then under the null hypothesis, by symmetry,

$$\mathbb{P}\left[Y_1 + Y_2 > 0, \ Y_1 + \tilde{Y}_2 > 0\right] = 1/3. \qquad (1.37)$$

As a consequence

$$\mathbb{V}(\tilde{f}_1) = \frac{1}{3} - \frac{1}{4} = \frac{1}{12} = \delta_1. \qquad (1.38)$$

Thus from the UCLT, Theorem 1.1 (since $m = 2$),

$$n^{1/2}(U_n(f) - 1/2) \xrightarrow{\mathcal{D}} N(0, 1/3).$$

Hence using (1.35), (1.36) and (1.38) we have,

$$n^{-3/2}\sqrt{12}(T^+ - n^2/4) \xrightarrow{\mathcal{D}} N(0, 1).$$

$\square$

1.4   Rate of convergence

The results of this section provide rate of convergence in the WLLN/SLLN and the UCLT Theorem 1.1. These results will be specifically useful for establishing rate of convergence results for M_m-estimators later in Chapter 2. Complete proofs of these results will not be provided here but the reader will be directed to other sources for full proofs.

The first result supplements the SLLN for U-statistics mentioned in Section 1.3. Parts (a) and (c) will also be useful in the proof of Theorem 2.4.

Theorem 1.5. *(Rate of convergence in SLLN.) Let $U_n(h)$ be a U-statistic with kernel h. Let $\mu = \mathbb{E}h(Y_1, \ldots, Y_m)$.*

(a) (Grams and Serfling (1973)). If $\mathbb{E}|h(Y_1, \ldots, Y_m)|^r < \infty$ for some $r > 1$, then for every $\epsilon > 0$,

$$\mathbb{P}\left[\sup_{k \geq n} |U_k(h) - \mu| > \epsilon\right] = o(n^{1-r}). \tag{1.39}$$

(b) If $\psi(s) = \mathbb{E}\left[\exp\left[s|h(Y_1, \ldots, Y_m)|\right]\right] < \infty$ for some $0 < s \leq s_0$, then for $k = [n/m]$, and $0 < s \leq s_0 k$,

$$\mathbb{E}\left[\exp(sU_n(h))\right] \leq \left[\psi(s/k)\right]^k. \tag{1.40}$$

(c) (Berk (1970)). Under the same assumption as (b), for every $\epsilon > 0$, there exist constants $0 < \delta < 1$, and C and such that

$$\mathbb{P}\left[\sup_{k \geq n} |U_k(h) - \mu| > \epsilon\right] \leq C\delta^n. \tag{1.41}$$

We provide a sketch of the proof of (a). First assume that $m = 1$. Then $U_n(h)$ reduces to a sample mean. In this case (a) is a direct consequence of the following well known result. A proof of this result can be found in Petrov (1975), Chapter 9, Theorem 2.8. We need the following lemma in order to establish (a).

Lemma 1.1 (Brillinger (1962); Wagner (1969)). *Suppose $Y_1, Y_2, \ldots$ are i.i.d. random variables with $\mathbb{E}|Y_n|^r < \infty$ for $r > 1$. Let $S_n = Y_1 + \ldots + Y_n$, and $\mu = \mathbb{E}Y_n$. Then for each $\varepsilon > 0$,*

$$\mathbb{P}\left[\sup_{k \geq n} \left|\frac{S_k}{k} - \mu\right| > \varepsilon\right] = o(n^{1-r}), \quad as \ n \to \infty. \tag{1.42}$$

Proof of Theorem 1.5: Now suppose $m > 1$. Recall the weak decomposition (1.28) of U-statistics and write

$$U_n(h) - \theta = \frac{m}{n} \sum_{i=1}^{n} \tilde{h}_1(Y_i) + R_n. \tag{1.43}$$

Since the result has already been established for $m = 1$, it is now enough to prove that

$$\mathbb{P}\left[\sup_{k \geq n} |R_k| \geq \epsilon\right] = o(n^{1-r}). \tag{1.44}$$

Note that since R_n is the sum of two U-statistics, it is also a U-statistic. Every U-statistic is a *reverse martingale* and from the well known reverse martingale inequality (see Shorack (2000), page 247) it follows that

$$\mathbb{P}\left[\sup_{k \geq n} |R_k| \geq \epsilon\right] \leq \epsilon^{-1}\mathbb{E}|R_n|. \tag{1.45}$$

Further, R_n is a *degenerate U-statistic*, that is, it is a U-statistic whose first projection is zero. Hence using Theorem 2.1.3 of Korolyuk and Borovskich (1993) (page 72), for $1 < r < 2$ and Theorem 2.1.4 of Korolyuk and Borovskich (1993) (page 73), for $r \geq 2$, it follows that

$$\mathbb{E}|R_n| = O(n^{2(1-r)}) = o(n^{1-r}). \tag{1.46}$$

Using this in (1.45) verifies (1.44) and proves Theorem 1.5(a) completely.

The detailed proofs of (b) and (c) (without the supremum) can be found in Serfling (1980) (page 200–202). We just mention that for the special case of $m = 1$, part (c) is an immediate consequence of the following lemma whose proof is available in Durrett (1991), Lemma 9.4, Chapter 1.   □

Lemma 1.2. *Let $Y_1, Y_2, \ldots$ be i.i.d. random variables, and $\mathbb{E}\left[e^{s|Y_n|}\right] < \infty$ for some $s > 0$. Let $S_n = Y_1 + \ldots + Y_n$, $\mu = \mathbb{E}Y_n$. Then for each $\varepsilon > 0$ there exists $a > 0$ such that*

$$\mathbb{P}\left[\left|\frac{S_n}{n} - \mu\right| > \varepsilon\right] = O(e^{-an}), \quad as \ n \to \infty. \tag{1.47}$$

The following result is on the *moderate deviation* of U-statistics. We allow the possibility that the kernel h is also changing with the sample size n. This result will be used in the proof of Theorem 2.5 of Chapter 2.

Theorem 1.6 (Bose (1997)). *(Deviation result) Let $\{h_n\}$ be a sequence of (symmetric) kernels of order m and let $\{Y_{ni}, 1 \leq i \leq n\}$ be i.i.d. real valued random variables for each n. Let*

$$U_n(h_n) = \binom{n}{m}^{-1} \sum_{1 \leq i_1 < \cdots < i_m \leq n} h_n(Y_{ni_1}, \ldots, Y_{ni_m}). \qquad (1.48)$$

Further suppose that for some $b, \delta > 0$, and some $v_n \leq n^\delta$,

$$\mathbb{E}\, h_n(Y_{ni_1}, \ldots, Y_{ni_m}) = 0,$$

$$\mathbb{E}\, |h_n(Y_{ni_1}, \ldots, Y_{ni_m})|^2 \leq v_n^2 \quad and$$

$$\mathbb{E}\, |h_n(Y_{ni_1}, \ldots, Y_{ni_m})|^r \leq b < \infty \quad for\ some\ r > 2.$$

Then for all large K,

$$\mathbb{P}\Big[n^{1/2}|U_n(h_n)| > Kv_n\,(\log n)^{1/2}\Big] \leq Dbn^{1-r/2}v_n^{-r}\,(\log n)^{r/2-1}. \qquad (1.49)$$

Proof of Theorem 1.6: Let

$$\widetilde{h}_n = h_n I(|h_n| \leq m_n), \quad h_{n1} = \widetilde{h}_n - \mathbb{E}\widetilde{h}_n, \quad h_{n2} = h_n - h_{n1},$$

where $\{m_n\}$ will be chosen.

Note that $\{h_{n1}\}$ and $\{h_{n2}\}$ are mean zero kernels and have the same properties as $\{h_n\}$ assumed in the statement of the theorem. Further

$$U_n(h_n) = U_n(h_{n1}) + U_n(h_{n2}).$$

Define for sufficiently large K,

$$a_n = K\,(\log n)^{1/2}/2 \quad and \quad \Psi_n(t) = \mathbb{E}[\exp\{tU_n(h_{n1}(Y_{n1}, \ldots, Y_{n,m}))\}].$$

Note that $\Psi_n(t)$ is finite for each t since h_{n1} is bounded. Letting $k = [n/m]$, and using Theorem 1.5(b),

$$
\begin{aligned}
A_1 &= \mathbb{P}(n^{1/2}U_n(h_{n1}) \geq v_n a_n) \\
&= \mathbb{P}(tn^{1/2}U_n(h_{n1})/v_n \geq t\,a_n) \\
&\leq \exp(-ta_n)\,\Psi_n(n^{1/2}t/v_n) \\
&\leq \exp(-ta_n)\,[\mathbb{E}\exp(n^{1/2}tY/(v_nk))]^k, \quad \text{say,}
\end{aligned}
$$

where $Y = h_{n1}(Y_{n1}, \cdots, Y_{nm})$. Using the fact that $|Y| \leq m_n$, $\mathbb{E}Y = 0$, and $\mathbb{E}Y^2 \leq v_n^2$, we get

$$\mathbb{E}\exp\left(\frac{n^{1/2}t}{kv_n}Y\right) \leq 1 + \mathbb{E}\sum_{j=2}^{\infty} Y^2 \left(\frac{n^{1/2}t}{kv_n}\right)^j m_n^{j-2}/j!$$

$$\leq 1 + \frac{t^2 n}{2k^2 v_n^2}(\mathbb{E}Y^2)\sum_{j=0}^{\infty}\left(\frac{n^{1/2}t}{kv_n}m_n\right)^j /j!$$

$$\leq 1 + \frac{t^2 n}{k^2}$$

provided $t \leq n^{-1/2}kv_n/(2m_n)$. With such a choice of t,

$$A_1 \leq \exp\left(-ta_n + \frac{t^2 n}{k}\right). \tag{1.50}$$

Let $t = K(\log n)^{1/2}/(4(2m-1))$. Then for all large n, the exponent in (1.50) equals

$$-\frac{K^2 \log n}{8(2m-1)} + \frac{nK^2 \log n}{16(2m-1)^2 k} \leq -\frac{K^2 \log n}{16(2m-1)}. \tag{1.51}$$

Thus we have shown that

$$\mathbb{P}(|n^{1/2}U_n(h_{n1})| > Kv_n(\log n)^{1/2}/2) \leq n^{-K^2/16(2m-1)}. \tag{1.52}$$

To tackle $U_n(h_{n2})$, we proceed as follows.

$$\mathbb{P}(|n^{1/2}U_n(h_{n2})| \geq a_n v_n/2) \leq 4v_n^{-1}a_n^{-1}n^{1/2}\mathbb{E}|h_{n2}(X_{n1}, \ldots, X_{nm})|$$

$$\leq 8v_n^{-1}a_n^{-1}n^{1/2}[\mathbb{E}|h_n|^r]^{1/r}[\mathbb{P}(|h_n| \geq m_n)]^{1-1/r}$$

$$\leq 8v_n^{-1}a_n^{-1}n^{1/2}b^{1/r}(m_n^{-r})^{1-1/r}b^{1-1/r}.$$

Choosing $m_n = n^{1/2}v_n/(K(\log n)^{1/2})$,

$$\mathbb{P}(|n^{1/2}U_n(h_{n2})| \geq a_n v_n/2) \leq 16bv_n^{-r}K^{r-2}n^{1-r/2}(\log n)^{r/2-1}. \tag{1.53}$$

Note that in the above proof the restrictions we have in place on m_n, K and t (with $t = K(\log n)^{1/2}/(4(2m-1)) \leq n^{-1/2}kv_n/(2m_n)$) are all compatible. The Theorem follows by using (1.52) and (1.53) and the given condition on v_n. $\square$

1.5   Degenerate U-statistics

In this section, we will consider distribution limits for degenerate U-statistics. Note that in Theorem 1.1 if $\delta_1 = V(h_1(Y_1)) = 0$, then the limit normal distribution is degenerate. That is, $n^{1/2}(U_n - \theta) \overset{\mathrm{P}}{\to} 0$. A natural question to ask is if we can renormalize U_n in cases where $\delta_1 = 0$ to obtain a non-degenerate limit.

Example 1.13: Consider the U-statistic $U_n = s_n^2$ from Example 1.9, which has the kernel

$$h(x_1, x_2) = \frac{(x_1 - x_2)^2}{2}.$$

Then as calculated earlier, $\delta_1 = \frac{1}{4}(\mu_4 - \mu_2^2)$ where

$$\mu_4 = \mathbb{E}(Y_1 - \mu)^4, \quad \mu_2 = \mathbb{E}(Y_1 - \mu)^2. \tag{1.54}$$

Note that

$$\mu_4 = \mu_2^2 \Leftrightarrow (Y_1 - \mu)^2 \text{ is a constant}$$
$$\Leftrightarrow Y_1 = \mu \pm C \ (C \text{ is a constant}).$$

Then $n^{1/2}(s_n^2 - \mu_2) \overset{\mathrm{P}}{\to} 0$. □

Example 1.14: Let f be a function such that $\mathbb{E}f(Y_1) = 0$ and $\mathbb{E}f^2(Y_1) < \infty$. Let U_n be the U-statistic with kernel $h(x_1, x_2) = f(x_1)f(x_2)$. Then $h_1(x) = \mathbb{E}f(x)f(Y_2) = 0$, and

$$n^{\frac{1}{2}}U_n(h) \overset{\mathcal{D}}{\longrightarrow} N(0, 0). \tag{1.55}$$

On the other hand,

$$U_n = \binom{n}{2}^{-1} \sum_{1 \leq i < j \leq n} f(Y_i)f(Y_j)$$
$$= \binom{n}{2}^{-1} \frac{1}{2} \sum_{i,j=1}^{n} f(Y_i)f(Y_j) - \frac{1}{n(n-1)} \sum_{i=1}^{n} f^2(Y_i)$$

$$= \frac{1}{n-1} \left(\frac{\sum_{i=1}^{n} f(Y_i)}{\sqrt{n}} \right)^2 - \frac{1}{n(n-1)} \sum_{i=1}^{n} f^2(Y_i).$$

As a consequence, using SLLN on $\{f^2(Y_i)\}$, and CLT on $\{f(Y_i)\}$,

$$nU_n \xrightarrow{\mathcal{D}} \sigma^2(\chi_1^2 - 1) \qquad (1.56)$$

where $\sigma^2 = \mathbb{E}f^2(Y_1)$ and χ_1^2 is a chi-square random variable with one degree of freedom.   □

Example 1.15: (continued from Example 1.13). Suppose $\{Y_i\}$'s are i.i.d.,

$$\mathbb{P}(Y_i = 1) = \mathbb{P}(Y_i = -1) = 1/2.$$

Then $\mu_4 = \mu_2^2 = 1$. Hence, as seen earlier,

$$n^{1/2}(s_n^2 - 1) \xrightarrow{\mathcal{D}} N(0,0). \qquad (1.57)$$

However, writing $\overline{Y} = n^{-1} \sum_{i=1}^{n} Y_i$ and noting that $Y_i^2 = 1$ for all i,

$$s_n^2 = \binom{n}{2}^{-1} \sum_{1 \le i < j \le n} \frac{(Y_i - Y_j)^2}{2}$$

$$= \frac{1}{n-1} \left(\sum_{i=1}^{n} Y_i^2 - n(\overline{Y})^2 \right)$$

$$= \frac{n}{n-1} - \frac{n}{n-1} \overline{Y}^2.$$

Hence $n \to \infty$

$$n(s_n^2 - 1) = \frac{n}{n-1} - (\sqrt{n}\overline{Y})^2 \frac{n}{n-1} \xrightarrow{\mathcal{D}} 1 - \chi_1^2.$$

□

Example 1.16: Suppose f_1 and f_2 are two functions such that

$$\mathbb{E}f_1(Y_1) = \mathbb{E}f_2(Y_1) = 0,$$

$\mathbb{E}(f_1(Y_1)f_2(Y_1)) = 0$ and $\mathbb{E}f_1^2(Y_1) = \mathbb{E}f_2^2(Y_2) = 1$. That is, $\{f_1, f_2\}$ are "orthonormal". Consider the U-statistic U_n with kernel

$$h(x_1, x_2) = a_1 f_1(x_1) f_1(x_2) + a_2 f_2(x_1) f_2(x_2). \tag{1.58}$$

Then as in Example 1.14, but now using the multivariate CLT,

$$nU_n \xrightarrow{\mathcal{D}} a_1(W_1 - 1) + a_2(W_2 - 1) \tag{1.59}$$

where W_1 and W_2 are i.i.d. χ_1^2 random variables. $\qquad\square$

To motivate further the limit result that we will state and prove shortly, let us continue to assume $\delta_1 = 0$. Recalling the formula for variance given in (1.16), now

$$\mathbb{V}(nU_n) = \frac{n^2}{\binom{n}{m}} \binom{m}{2} \binom{n-m}{m-2} \delta_2 + \text{ smaller order terms}$$

$$= \frac{[m(m-1)]^2}{2} \delta_2 + o(1).$$

Thus if $\delta_1 = 0$, then the right scaling is indeed nU_n. To understand further our forthcoming result, let

$$L_2(\mathbb{F}) = \left\{ h : \mathbb{R} \to \mathbb{R}, \int h^2(x) d\mathbb{F}(x) < \infty \right\} \text{ and} \tag{1.60}$$

$$L_2(\mathbb{F} \times \mathbb{F}) = \left\{ h : \mathbb{R} \times \mathbb{R} \to \mathbb{R}, \int\int h^2(x, y) d\mathbb{F}(x) d\mathbb{F}(y) < \infty \right\}. \tag{1.61}$$

Fix $h \in L_2(\mathbb{F} \times \mathbb{F})$. Define the operator $A_h : L_2(\mathbb{F}) \to L_2(\mathbb{F})$ as

$$A_h g(x) = \int h(x, y) g(y) d\mathbb{F}(y), \; g \in L_2(\mathbb{F}). \tag{1.62}$$

Then there exists eigenvalues $\{\lambda_j\}$ and corresponding eigenfunctions $\{\phi_j\} \subset L_2(\mathbb{F})$ for the operator A. That is:

$$A_h \phi_j = \lambda_j \phi_j, \forall j, \tag{1.63}$$

$$\int \phi_j^2(x) d\mathbb{F}(x) = 1, \; \int \phi_j(x) \phi_k(x) d\mathbb{F}(x) = 0, \; \forall j \neq k, \text{ and} \tag{1.64}$$

$$h(x, y) = \sum_{j=1}^{\infty} \lambda_j \phi_j(x) \phi_j(y). \tag{1.65}$$

The equality in (1.65) is in the L_2 sense. That is, if Y_1, Y_2 are i.i.d. F then

$$\mathbb{E}[h(Y_1, Y_2) - \sum_{k=1}^{n} \lambda_k \phi_k(Y_1)\phi_k(Y_2)]^2 \to 0 \quad \text{as} \quad n \to \infty. \qquad (1.66)$$

Further

$$h_1(x) = \mathbb{E}h(x, Y_2)$$
$$= \sum_{k=1}^{\infty} \lambda_k \phi_k(x)\mathbb{E}(\phi_k(Y_2)) \quad \text{almost surely} \quad \mathbb{F}. \qquad (1.67)$$

Also note that from (1.64) and (1.66),

$$\mathbb{E}h^2(Y_1, Y_2) = \sum_{k=1}^{\infty} \lambda_k^2. \qquad (1.68)$$

Now we are ready to state our first theorem on degenerate U-statistics for $m = 2$. The version of this Theorem for $m > 2$ is given later in Theorem 1.9.

Theorem 1.7 (Gregory (1977); Serfling (1980)). *(χ^2-limit theorem.) Suppose $h(x,y)$ is a kernel such that $\mathbb{E}h(x, Y_1) = 0$ a.e. x and $\mathbb{E}h^2(Y_1, Y_2) < \infty$. Then*

$$nU_n \xrightarrow{\mathcal{D}} \sum_{k=1}^{\infty} \lambda_k(W_k - 1) \qquad (1.69)$$

where $\{W_k\}$ are i.i.d. χ_1^2 random variables and $\{\lambda_k\}$ are the (non-zero) eigenvalues of the operator A_h given in (1.62).

Proof of Theorem 1.7: The idea of the proof is really as in Example 1.16 after reducing the infinitely many eigenvalues case to the finitely many eigenvalues case. First note that

$$h_1(x) = \mathbb{E}h(x, Y_1) = 0 \quad \text{a.e.} \quad \mathbb{F}. \qquad (1.70)$$

On the other hand from (1.67), we know that

$$h_1(x) = \sum_{k=1}^{\infty} \lambda_k \phi_k(x)\mathbb{E}_F(\phi_k(Y_1)) \quad \text{a.e.} \quad \mathbb{F}. \qquad (1.71)$$

Since $\{\phi_k\}$ is an orthonormal system, using (1.70) we thus have

$$\mathbb{E}\phi_k(Y_1) = 0 \quad \forall k. \tag{1.72}$$

Also recall that (see (1.65))

$$h(x,y) = \sum_{k=1}^{\infty} \lambda_k \phi_k(x)\phi_k(y) \quad \text{in the} \quad L_2(\mathbb{F} \times \mathbb{F}) \quad \text{sense.} \tag{1.73}$$

Now

$$\begin{aligned}
nU_n &= n\binom{n}{2}^{-1} \sum_{1 \le i < j \le n} h(Y_i, Y_j) \\
&= \frac{1}{n-1} \sum_{1 \le i \ne j \le n} h(Y_i, Y_j).
\end{aligned}$$

Thus it is enough to find the limit distribution of

$$\begin{aligned}
T_n &= \frac{1}{n} \sum_{1 \le i \ne j \le n} h(Y_i, Y_j) \\
&= \frac{1}{n} \sum_{1 \le i \ne j \le n} \sum_{k=1}^{\infty} \lambda_k \phi_k(Y_i)\phi_k(Y_j). \tag{1.74}
\end{aligned}$$

If the sum over k were a finite sum then the rest of the proof would proceed exactly as in Example 1.16. With this in mind, define the truncated sum

$$T_{nk} = \frac{1}{n} \sum_{1 \le i \ne j \le n} \sum_{t=1}^{k} \lambda_t \phi_t(Y_i)\phi_t(Y_j) \quad k \ge 1. \tag{1.75}$$

Lemma 1.3. *Suppose for every k, $\{T_{nk}\}$ is any sequence of random variables and $\{T_n\}$ is another sequence of random variables, all on the same probability space, such that*

(i) $T_{nk} \xrightarrow{D} A_k$ as $n \to \infty$

(ii) $A_k \xrightarrow{D} A$ as $k \to \infty$

(iii) $\displaystyle\lim_{k\to\infty} \limsup_{n\to\infty} \mathbb{P}(|T_n - T_{nk}| > \epsilon) = 0 \ \forall\, \epsilon > 0$.

Then $T_n \xrightarrow{D} A$.

Proof of Lemma 1.3. Let $\phi_X(t) = \mathbb{E}(e^{itX})$ denote the characteristic function of any random variable X. It suffices to show that

$$\phi_{T_n}(t) \to \phi_A(t) \quad \forall t \in \mathbb{R}.$$

Now

$$|\phi_{T_n}(t) - \phi_A(t)| \leq |\phi_{T_n}(t) - \phi_{T_{nk}}(t)| + |\phi_{T_{nk}}(t) - \phi_{A_k}(t)|$$
$$+ |\phi_{A_k}(t) - \phi_A(t)|$$
$$= B_1 + B_2 + B_3, \text{ say.}$$

First let $n \to \infty$. Then by (i), $B_2 \to 0$. Now let $k \to \infty$. Then by (ii), $B_3 \to 0$. Now we tackle the first term B_1.

Fix $\epsilon > 0$. It is easy to see that

$$|\phi_{T_n}(t) - \phi_{T_{nk}}(t)| \leq \mathbb{E}[|e^{it(T_n - T_{nk})} - 1|]$$
$$\leq 2\mathbb{P}(|T_n - T_{nk}| > \epsilon) + \sup_{|x| \leq |t|\epsilon} |e^{ix} - 1|.$$

Now first let $n \to \infty$, then let $k \to \infty$ and use condition (iii) to conclude that the first term goes to zero. Now let $\epsilon \to 0$ to conclude that the second term goes to zero. Thus $B_1 \to 0$. This proves the lemma. $\square$

Now we continue with the proof of the Theorem. We shall apply Lemma 1.3 to $\{T_n\}$ and $\{T_{nk}\}$ defined in (1.74) and (1.75). Suppose $\{W_k\}$ is a sequence of i.i.d. χ_1^2 random variables.

Let

$$A_k = \sum_{j=1}^{k} \lambda_j(W_j - 1), \quad A = \sum_{j=1}^{\infty} \lambda_j(W_j - 1). \tag{1.76}$$

Note that by Kolmogorov's three series theorem (see for example Chow and Teicher (1997) Corollary 3, page 117), the value of the infinite series in (1.76) is finite almost surely, and hence A is a legitimate random variable.

Also note that $A_k \xrightarrow{D} A$ as $k \to \infty$. Hence, condition (ii) of Lemma 1.3 is verified. Now we claim that $T_{nk} \xrightarrow{D} A_k$. This is because

$$T_{nk} = \frac{1}{n} \sum_{1 \leq i \neq j \leq n} \sum_{t=1}^{k} \lambda_t \, \phi_t(Y_i)\phi_t(Y_j)$$
$$= \frac{1}{n} \sum_{t=1}^{k} \lambda_t \left(\sum_{i=1}^{n} \phi_t(Y_i)\right)^2 - \frac{1}{n} \sum_{t=1}^{k} \lambda_t \sum_{i=1}^{n} \phi_t^2(Y_i). \tag{1.77}$$

By SLLN, for every t,

$$\frac{1}{n} \sum_{i=1}^{n} \phi_t^2(Y_i) \xrightarrow{a.s.} \mathbb{E}\phi_t^2(Y_1) = 1. \tag{1.78}$$

By multivariate CLT, since $\{\phi_k(\cdot)\}$ are orthonormal,

$$\left(\frac{1}{\sqrt{n}} \sum_{i=1}^{n} \phi_t(Y_i), t = 1 \ldots k\right) \xrightarrow{D} N\left(0, \mathbb{I}_k\right) \tag{1.79}$$

where $\mathbb{I}_k$ is the $k \times k$ identity matrix. As a consequence, using (1.77), (1.78) and (1.79),

$$T_{nk} \xrightarrow{D} \sum_{t=1}^{k} \lambda_t W_t - \sum_{t=1}^{k} \lambda_t = \sum_{t=1}^{k} \lambda_t (W_t - 1) \tag{1.80}$$

where $\{W_i\}$ are i.i.d. χ_1^2 random variables. This verifies condition (i). To verify condition (iii) of Lemma 1.3, consider

$$
\begin{aligned}
T_n - T_{nk} &= \frac{1}{n} \sum_{1 \le i \ne j \le n} h(Y_i, Y_j) - \frac{1}{n} \sum_{1 \le i \ne j \le n} \sum_{t=1}^{k} \lambda_t \phi_t(Y_i)\phi_t(Y_j) \\
&= \left(\frac{2}{n}\right)\binom{n}{2} U_{nk}
\end{aligned}
\tag{1.81}
$$

where U_{nk} is a U-statistic of degree 2 with kernel

$$g_k(x, y) = h(x, y) - \sum_{t=1}^{k} \lambda_k \phi_k(x)\phi_k(y).$$

Now since h is a degenerate kernel and $\mathbb{E}\phi_k(Y_1) = 0 \; \forall k$, we conclude that g_k is also a degenerate kernel. Note that using (1.73)

$$
\begin{aligned}
\mathbb{E}g_k^2(Y_1, Y_2) &= \mathbb{E}[h(Y_1, Y_2) - \sum_{t=1}^{k} \lambda_t \phi_t(Y_1)\phi_t(Y_2)]^2 \\
&= \mathbb{E}\left[\sum_{t=k+1}^{\infty} \lambda_t \phi_t(Y_1)\phi_t(Y_2)\right]^2 \\
&= \sum_{t=k+1}^{\infty} \lambda_t^2.
\end{aligned}
$$

Hence from (1.81) using formula (1.16) for variance of a U-statistic,

$$\mathbb{E}(T_n - T_{nk})^2 = (n-1)^2 \mathbb{V}U_{nk}$$

$$= \frac{(n-1)^2}{\binom{n}{2}} \sum_{t=k+1}^{\infty} \lambda_t^2 \leq 2 \sum_{t=k+1}^{\infty} \lambda_t^2. \tag{1.82}$$

Hence

$$\lim_{k\to\infty} \limsup_{n\to\infty} \mathbb{P}(|T_n - T_{nk}| > \epsilon) \leq \frac{1}{\epsilon^2} \lim_{k\to\infty} \sum_{t=k+1}^{\infty} \lambda_t^2 = 0.$$

This establishes condition (iii) of Lemma 1.3 and hence the Theorem is completely proved. □

Before we provide an application of Theorem 1.7, we define the following:

Definition 1.2: For any sequence of numbers $\{x_1, \ldots x_n\}$, its empirical cumulative distribution function (e.c.d.f.) is defined as

$$\mathbb{F}_n(x) = \frac{1}{n} \sum_{i=1}^{n} \mathcal{I}_{\{x_i \leq x\}}. \tag{1.83}$$

Example 1.17: *(Cramér-von Mises statistic):* Let $Y_1, \ldots, Y_n$ be i.i.d. with distribution function $\mathbb{F}$. The Cramér-von Mises statistic is often used to test the null hypothesis that F equals a specified c.d.f. $\mathbb{F}_0$. It is defined as

$$CV_n = n \int \big(\mathbb{F}_n(x) - \mathbb{F}_0(x)\big)^2 d\mathbb{F}_0(x) \tag{1.84}$$

where $\mathbb{F}_n(\cdot)$ is the e.c.d.f. of $\{Y_1, \ldots, Y_n\}$. It is not hard to see that if $\mathbb{F}_0$ is continuous, then under the null hypothesis, CV_n is *distribution free*. That is, its distribution is independent of $\mathbb{F}_0$ whenever $Y_1, \ldots, Y_n$ are i.i.d. $\mathbb{F}_0$. We shall see below that CV_n is a degenerate U-statistic. The following result establishes the asymptotic distribution of CV_n. □

Theorem 1.8. *If $\mathbb{F}_0$ is continuous and $Y_1, Y_2, \ldots$, are i.i.d. with c.d.f. $\mathbb{F}_0$ then with $\{W_k\}$ as i.i.d. χ_1^2 random variables,*

$$CV_n \xrightarrow{\mathcal{D}} \frac{1}{\pi^2} \sum_{k=1}^{\infty} \frac{W_k}{k^2}. \tag{1.85}$$

Proof of Theorem 1.8: Since the distribution of CV_n is independent of $\mathbb{F}_0$, without loss of generality suppose that $\mathbb{F}_0$ is the c.d.f. of the uniform distribution:

$$\mathbb{F}_0(x) = x, \quad 0 \leq x \leq 1 \tag{1.86}$$

and hence

$$
\begin{aligned}
CV_n &= n \int_0^1 (\mathbb{F}_n(x) - x)^2 \, dx \\
&= n \int_0^1 \left[\frac{1}{n} \sum_{i=1}^n (\mathcal{I}_{\{Y_i \leq x\}} - x) \right]^2 dx \tag{1.87} \\
&= \frac{n}{n^2} \sum_{1 \leq i,j \leq n} \int_0^1 (\mathcal{I}_{\{Y_i \leq x\}} - x)(\mathcal{I}_{\{Y_j \leq x\}} - x) \, dx \\
&= \frac{2}{n} \sum_{1 \leq i < j \leq n} \int_0^1 (\mathcal{I}_{\{Y_i \leq x\}} - x)(\mathcal{I}_{\{Y_j \leq x\}} - x) \, dx \\
&\quad + \frac{1}{n} \sum_{i=1}^n \int_0^1 (\mathcal{I}_{\{Y_i \leq x\}} - x)^2 \, dx \\
&= \binom{n}{2} \left(\frac{2}{n} \right) U_n(f) + U_n(h). \tag{1.88}
\end{aligned}
$$

Here $U_n(f)$ and $U_n(h)$ are U-statistics with kernels respectively,

$$f(x_1, x_2) = \int_0^1 (\mathcal{I}_{\{x_1 \leq x\}} - x)(\mathcal{I}_{\{x_2 \leq x\}} - x) \, dx,$$

$$h(x_1) = \int_0^1 (\mathcal{I}_{\{x_1 \leq x\}} - x)^2 \, dx.$$

Note that $\mathcal{I}_{\{Y_i \leq x\}}$ are i.i.d. Bernoulli random variables with probability of success x. Hence by SLLN

$$U_n(h) \overset{a.s.}{\to} \mathbb{E} \int_0^1 (\mathcal{I}_{\{Y_i \leq x\}} - x)^2 \, dx \tag{1.89}$$

$$= \int_0^1 \mathbb{E}(\mathcal{I}_{\{Y_i \leq x\}} - x)^2 \, dx \tag{1.90}$$

$$= \int_0^1 x(1 - x) \, dx = \frac{1}{6}. \tag{1.91}$$

Moreover

$$\mathbb{E}f(x_1, Y_2) = \mathbb{E}\int_0^1 \left(\mathcal{I}_{\{x_1 \le x\}} - x\right)\left(\mathcal{I}_{\{Y_2 \le x\}} - x\right)dx,$$

$$= 0.$$

Thus $U_n(f)$ is a degenerate U-statistic. Hence by Theorem 1.7,

$$nU_n(f) \xrightarrow{\mathcal{D}} \sum_{k=1}^\infty \lambda_k(W_k - 1) \tag{1.92}$$

where $\{W_k\}$ are i.i.d. χ_1^2 variables and $\{\lambda_k\}$ are the eigenvalues of the kernel f. We now identify the values $\{\lambda_k\}$. The eigenequation is

$$\int_0^1 f(x_1, x_2)g(x_2)dx_2 = \lambda g(x_1). \tag{1.93}$$

Now

$$f(x_1, x_2) = \int_0^1 \left[\mathcal{I}(x_1 \le x)\mathcal{I}(x_2 \le x) - x\,\mathcal{I}(x_1 \le x) - x\,\mathcal{I}(x_2 \le x) + x^2\right]dx$$

$$= \int_0^1 \left[\mathcal{I}(x \ge \max(x_1, x_2)) - x\,\mathcal{I}(x_1 \le x) - x\,\mathcal{I}(x_2 \le x) + x^2\right]dx$$

$$= 1 - \max(x_1, x_2) - \frac{1 - x_1^2}{2} - \frac{1 - x_2^2}{2} + \frac{1}{3}$$

$$= \frac{1}{3} - \max(x_1, x_2) + \frac{x_1^2 + x_2^2}{2}. \tag{1.94}$$

Recall that any eigenfunction g must satisfy $\int_0^1 g(x)dx = 0$ (see (1.72)). Hence using (1.94), (1.93) reduces to

$$\lambda g(x_1) = \int_0^1 \left(\frac{1}{3} - \max(x_1, x_2) + \frac{x_1^2 + x_2^2}{2}\right)g(x_2)dx_2$$

$$= \int_0^1 \frac{x_2^2 g(x_2)}{2}dx_2 - \int_{x_1}^1 x_2 g(x_2)dx_2 - x_1\int_0^{x_1} g(x_2)dx_2. \tag{1.95}$$

For the moment, assume that g is a continuous function. Then (1.95) implies that g is differentiable and hence taking derivative w.r.t. x_1,

$$
\begin{aligned}
\lambda g'(x_1) &= x_1 g(x_1) - \int_0^{x_1} g(x_2)dx_2 - x_1 g(x_1) \\
&= -\int_0^{x_1} g(x_2)dx_2.
\end{aligned}
$$

Again, if g is continuous, then the right side of the above is differentiable, and hence so is the left side. Thus we get

$$
\lambda g''(x_1) = -g(x_1). \tag{1.96}
$$

It is well known that the general solution to this second order differential equation is

$$
g(x) = C_1 e^{itx} + C_2 e^{-itx}. \tag{1.97}
$$

The boundary conditions are

$$
\int_0^1 g(x)dx = 0 \quad \text{and} \quad \int_0^1 g^2(x)dx = 1. \tag{1.98}
$$

Since g cannot be the constant function, $t \neq 0$. Since g is real, $C_2 = C_1$. Thus

$$
\begin{aligned}
g(x) &= C_1(e^{itx} + e^{-itx}) \\
&= 2C_1 \cos(tx), \quad 0 \le x \le 1.
\end{aligned}
$$

Now

$$
0 = \int_0^1 g(x)dx \Rightarrow t = \pi k, \ k = 0, \pm 1 \ldots \tag{1.99}
$$

Further

$$
\begin{aligned}
1 &= \int_0^1 g^2(x)dx \\
&= 4C_1^2 \int_0^1 \cos^2(\pi k x)dx \\
&= \frac{4C_1^2}{2} = 2C_1^2.
\end{aligned}
$$

Now using (1.96),

$$-\lambda(2C_1)t^2\cos(tx) = -2C_1\cos(tx) \qquad (1.100)$$

which implies $\lambda t^2 = 1$.

As a consequence, since $t = \pi k$,

$$\lambda\pi^2 k^2 = 1, \quad k = \pm 1\ldots \qquad (1.101)$$

or

$$\lambda = \frac{1}{\pi^2 k^2}, \quad k = 1, 2, \ldots \qquad (1.102)$$

(with possible multiplicity). But $\{\sqrt{2}\cos(\pi kx), \quad k = 1, 2, \ldots\}$ is a complete orthonormal system and hence we can conclude that the eigenvalues (with no multiplicities) and the eigenfunctions are given by

$$\left\{(\lambda_k = \frac{1}{\pi^2 k^2}, \ g_k(x) = \sqrt{2}\ \cos(\pi kx)), \quad k = 1, \ldots\right\}. \qquad (1.103)$$

Notice that $\sum_{k=1}^{\infty}\lambda_k = 1/6$ (which was first proved by Euler, see Knopp (1923) for an early proof).

As a consequence, using (1.88), (1.91) and (1.92),

$$CV_n \xrightarrow{\mathcal{D}} \sum_{k=1}^{\infty}\lambda_k(W_k - 1) + \frac{1}{6} = \sum\lambda_k W_k = \frac{1}{\pi^2}\sum_{k=1}^{\infty}\frac{W_k}{k^2}\ .$$

$\square$

Example 1.18: (Example 1.7 continued) Using (1.13), it is easy to see that κ_n is a degenerate U-statistic. An application of Theorem 1.7 yields the following: suppose $\{(X_i, Y_i)\}$ are i.i.d. and moreover all the $\{X_i, Y_j\}$ are independent. Further suppose that their second moments are finite. Then

$$n\kappa_n \xrightarrow{\mathcal{D}} \sum_{i,j=1}^{\infty}\lambda_i\mu_j Z_{i,j}$$

where $\{Z_{i,j}\}$ are i.i.d. chi-square random variables each with one degree of freedom. The $\{\lambda_i\}$ and $\{\mu_j\}$ are given by the solution of the eigenvalue equations:

$$\int h_{F_X}(x_1, x_2)g_X(x_2)dF_X(x_2) = \lambda g_X(x_1) \quad a.e.(F_X). \tag{1.104}$$

$$\int h_{F_Y}(y_1, y_2)g_Y(y_2)dF_Y(y_2) = \lambda g_Y(y_1) \quad a.e.(F_Y). \tag{1.105}$$

$\square$

We now state the asymptotic limit law for degenerate U-statistics for general $m > 2$. Define the *second projection* h_2 of the kernel h as

$$h_2(x_1, x_2) = \mathbb{E}h_2(x_1, x_2, Y_3, \ldots, Y_m).$$

Note that h_2 is a symmetric kernel of order 2.

Theorem 1.9. *Suppose* $Y_1, \ldots, Y_n$ *are i.i.d.* $\mathbb{F}$ *and* U_n *is a* U-*statistic with kernel* h *of order* $m \geq 2$ *such that* $\mathbb{E}h(x, Y_2, \ldots, Y_m) = 0$ *almost surely and* $\mathbb{E}h^2(Y_1, Y_2, \ldots, Y_m) < \infty$. *Then*

$$nU_n \xrightarrow{\mathcal{D}} \binom{m}{2} \sum_{k=1}^{\infty} \lambda_k(W_k - 1) \tag{1.106}$$

where $\{W_k\}$ *are i.i.d.* χ_1^2 *random variables and* $\{\lambda_k\}$ *are the eigenvalues of the operator defined by*

$$Ag(x_2) = \int h_2(x_1, x_2)g(x_1)d\mathbb{F}(x_1).$$

We omit the proof of this Theorem, which is easy once we use Theorem 1.7 and some easily derivable properties of the second-order remainder in the Hoeffding decomposition. For details see Lee (1990), page 83.

Before concluding, we also note that in a remarkable article, Dynkin and Mandelbaum (1983) study the limit distribution of symmetric statistics in a unified way. This in particular provide the limit distribution of U-statistics, both degenerate and non-degenerate. Unfortunately, this requires ideas from Poisson point processes and Wiener integrals, which are outside the purview of this book.

1.6   Exercises

1. Suppose $\mathcal{F}$ is the set of all cumulative distribution functions on $\mathbb{R}$. Let $Y_1, \ldots, Y_n$ be i.i.d. observations from some unknown $\mathbb{F} \in \mathcal{F}$.

    (a) Show that the e.c.d.f. $\mathbb{F}_n$ is a complete sufficient statistic for this space.

    (b) Show that the e.c.d.f. $\mathbb{F}_n$ is the nonparametric maximum likelihood estimator of the unknown c.d.f. $\mathbb{F}$.

    (c) Using the above fact, show that any U-statistic (with finite variance) is the nonparametric UMVUE of its expectation.

2. Verify (1.4), (1.5) and (1.8).

3. Justify that Gini's mean difference is a measure of inequality.

4. Show that any U-statistic of degree m is also a U-statistic of degree $(m+1)$.

5. Show that for the U-statistic Kendall's tau given in Example 1.4, under independence of X and Y,

$$\delta_1 = 1/9, \qquad \delta_2 = 1, \text{ and hence}$$

$$\mathbb{V}U_n = \frac{2(2n+5)}{9n(n+1)} = \frac{4}{9n} + O(n^{-2}).$$

6. Suppose $U_n(f)$ and $U_n(g)$ are two U-statistics with kernels of order m_1 and m_2 respectively, $m_1 \leq m_2$. Generalize formula (1.16) by showing that

$$\mathbb{COV}(U_n(f), U_n(g)) = \binom{n}{m_1}^{-1} \sum_{c=1}^{m_1} \binom{m_2}{c} \binom{n-m_2}{m_2-c} \delta_{c,c}^2$$

where

$$\delta_{c,c}^2 = \mathbb{COV}\big(f(X_{i_1}, \ldots, X_{i_{m_1}}), g(X_{j_1}, \ldots, X_{j_{m_2}})\big)$$

and c = number of common indices between the sets $\{i_1, \ldots, i_{m_1}\}$ and $\{j_1, \ldots, j_{m_2}\}$.

7. Using the previous exercise, show that for Wilcoxon's T^+ given in Example 1.6,

$$V(T^+) = \binom{n}{2}\left[(n-1)(p_4 - p_2^2) + p_2(1 - p_2) + 4(p_3 - p_1 p_2)\right] + np_1(1 - p_1)$$

where

$$p_1 = \mathbb{P}(Y_1 > 0), \quad p_2 = \mathbb{P}(Y_1 + Y_2 > 0),$$
$$p_3 = \mathbb{P}(Y_1 + Y_2 > 0), \quad p_4 = \mathbb{P}(Y_1 + Y_2 > 0, Y_2 + Y_3 > 0).$$

Further, if the distribution of Y_1 is symmetric about zero, then $p_1 = 1/2$, $p_2 = 1/2$, $p_3 = 3/8$, and $p_4 = 1/3$. As a consequence

$$\mathbb{E}T^+ = \frac{n(n+1)}{4} \quad \text{and} \quad \mathbb{V}T^+ = \frac{n(n+1)(2n+1)}{24}.$$

8. Verify (1.18) for the limiting variance of a U-statistic.

9. Verify (1.20) for sample variance.

10. Show that for the sample covariance U-statistics given in Example 1.3,

$$\delta_1 = (\mu_{2,2} - \sigma_{X,Y}^2)/4,$$
$$\delta_2 = (\mu_{2,2} + \sigma_X^2 \sigma_Y^2)/2, \quad \text{and hence}$$
$$\mathbb{V}U_n = n^{-1}\mu_{2,2} - \frac{(n-2)\sigma_{XY}^2 - \sigma_X^2 \sigma_Y^2}{n(n-1)},$$

where

$$\mu_{2,2} = \mathbb{E}[(X - \mu_X)^2 (Y - \mu_Y)^2]$$
$$\sigma_X^2 = \mathbb{V}(X), \quad \sigma_Y^2 = \mathbb{V}(Y),$$
$$\sigma_{XY}^2 = \mathbb{COV}(X, Y).$$

11. Verify the orthogonal decomposition relation (1.23).

12. Verify (1.30) for the first projection of Kendall's tau.

13. Let $Y_{(1)} \leq \cdots \leq Y_{(n)}$ be the order statistics of $Y_1, \ldots, Y_n$. Derive a connection between the Gini's mean difference given in Example 1.5

and the L-statistic

$$L_n = \sum_{i=1}^{n} iY_{(i)} \qquad (1.107)$$

and thereby prove an asymptotic normality result for L_n.

14. Formulate and prove a multivariate version of Theorem 1.1.

15. Show that R_n defined in (1.22) is a degenerate U-statistic.

16. Consider the U-statistic U_n with kernel $h(x_1, x_2, x_3) = x_1 x_2 x_3$ where $\mathbb{E}Y_i = 0, \mathbb{V}(Y_i) = 1$. Show that $n^{3/2} U_n \xrightarrow{D} Z^3 - 3Z$ where $Z \sim N(0, 1)$.

17. Let U_n be the U-statistic with kernel $h(x_1, x_2, x_3, x_4) = x_1 x_2 x_3 x_4$ and $\mathbb{E}Y_1 = 0, \mathbb{V}(Y_1) = 1$. Show that $n^2 U_n \xrightarrow{D} Z^4 - 6Z^2 + 3$ where $Z \sim N(0, 1)$.

18. Verify (1.33) that U and V are independent continuous $U(-1, 1)$ random variables.

19. Show that in case of the sample variance in the degenerate case, there is only one eigenvalue equal to -1.

20. Show that (1.72) holds.

21. Suppose $Y_1, \ldots, Y_n$ are i.i.d. from $\mathbb{F}_0$ which is continuous. Show that $\mathbb{E}CV_n = 1/6$ for all $n \geq 1$ and its distribution does not depend on $\mathbb{F}_0$.

22. Suppose $Y \sim \mathbb{F}$ is such that $\mathbb{E}Y = 0, \mathbb{E}Y^2 = \sigma^2, \mathbb{E}Y^3 = 0$ and $\mathbb{E}Y^4 < \infty$. Consider the kernel

$$h(x, y) = xy + (x^2 - \sigma^2)(y^2 - \sigma^2).$$

Show that $U_n(h)$ is degenerate and the L_2 operator A_h has two eigenvalues and eigenfunctions. Find these and the asymptotic distribution of $nU_n(h)$.

23. Suppose $Y \sim \mathbb{F}$ is such that $\mathbb{E}Y = \mathbb{E}Y^3 = 0$, and $\mathbb{E}Y^6 < \infty$. Consider the kernel

$$h(x, y) = xy + x^3 y^3.$$

Show that $U_n(h)$ is degenerate. Find the eigenvalues and eigenfunctions of the operator A_h. Find the asymptotic distribution of $nU_n(h)$.

24. Suppose $Y \sim \mathbb{F}$ is an $\mathbb{R}^d$ valued random variable such that Y and $-Y$ have the same distribution. Consider the kernel

$$h(x, y) = |x + y| - |x - y|,$$

where $|a| = \left(\sum_{j=1}^d a_j^2\right)^{1/2}$ for $a = (a_1, \ldots, a_d) \in \mathbb{R}^d$. Show that $U_n(h)$ is a degenerate U-statistic. Find the asymptotic distribution of $nU_n(h)$ when $d = 1$.

25. (This result is used in the proof of Theorem 2.3.) Recall the variance formula (1.18).

    (a) Show that

    $$\mathbb{V}\Big[h_1(Y_1)\Big] = \delta_1 \leq \delta_2 \leq \cdots \leq \delta_m = \mathbb{V}\Big[h(Y_1, \ldots, Y_m)\Big].$$

    (b) Show that

    $$\mathbb{V}\Big[U_n\Big] \leq \frac{m}{n} \delta_m.$$

26. For Bergsma's κ in Example 1.18, show that the eigenvalues and eigenfunctions when all the random variables are i.i.d. continuous uniform $(0, 1)$ are given by $\{\frac{1}{\pi k^2}, g_k(u) = 2\cos(\pi k u), 0 \leq u \leq 1, k = 1, 2, \ldots\}$.

27. To have an idea of how fast the asymptotic distribution takes hold in the degenerate case, plot $(k, \sum_{i=1}^k \lambda_i / \sum_{i=1}^\infty \lambda_i), k = 1, 2, \ldots$ for (a) the Cramer-von Mises statistic (b) Bergma's κ_n, when all distributions are continuous uniform $(0, 1)$.

Chapter 2

M_m-estimators and U-statistics

M-estimators, and their general versions M_m-estimators, were introduced by
Huber (1964) out of robustness considerations. The literature on these estimators is very rich and the asymptotic properties of these estimates have
been treated under different sets of conditions. To establish the most general
results for these estimators require very sophisticated treatment using techniques from the theory of empirical processes. But here we strive for a simple
approach.

The goal of this chapter is to first establish a link between U-statistics and
M_m-estimators by imposing a few simple conditions, including a convexity
condition. Yet, the conditions are general enough to be applicable widely. A
huge class of M_m-estimators turn out to be approximate U-statistics. Hence
the theory of U-statistics can be used to derive asymptotic properties of M_m-
estimators. We give several examples to show how the general results can
be applied to many specific estimators. In particular, several multivariate
estimates of location are discussed in details.

2.1 Basic definitions and examples

Let $f(x_1, \ldots, x_m, \theta)$ be a real valued measurable function which is symmetric
in the arguments $x_1, \ldots, x_m$. The argument θ is assumed to belong to $\mathbb{R}^d$
and x_i's belong to some measure space $\mathcal{Y}$.

© Springer Nature Singapore Pte Ltd. 2018 and Hindustan Book Agency 2018
A. Bose and S. Chatterjee, *U-Statistics, Mm-Estimators and Resampling*, Texts
and Readings in Mathematics 75, https://doi.org/10.1007/978-981-13-2248-8_2

Definition 2.1 *(M_m-parameter)*: Let $Y_1, \ldots, Y_m$ be i.i.d. $\mathcal{Y}$-valued random variables. Define

$$Q(\theta) = \mathbb{E}\, f(Y_1, \ldots, Y_m, \theta). \tag{2.1}$$

Let θ_0 be the minimizer of $Q(\theta)$. Assume that θ_0 is unique. We consider θ_0 to be the unknown parameter. It is called an M_m-*parameter.*

The special case when $m = 1$ is the one that is most commonly studied and in that case θ_0 is traditionally called an M-*parameter.*

The sample analogue Q_n of Q is given by

$$Q_n(\theta) = \binom{n}{m}^{-1} \sum_{1 \leq i_1 < i_2 \ldots < i_m \leq n} f(Y_{i_1}, \ldots, Y_{i_m}, \theta). \tag{2.2}$$

In the absence of any further information on the distribution of Y_i, a natural (nonparametric) estimate of the M_m-parameter θ_0 is the minimizer of $Q_n(\theta)$.

Definition 2.2 *(M_m-estimator)*: Suppose that $\{Y_1, \cdots, Y_n\}$ is a sequence of i.i.d. observations. Any measurable value θ_n which minimizes $Q_n(\theta)$ is called an M_m-*estimator* of θ_0.

So

$$Q_n(\theta_n) = \inf Q_n(\theta). \tag{2.3}$$

Note that even when θ_0 is unique, θ_n need not be unique. By an appropriate *Selection Theorem* it is often possible to choose a measurable version. We shall see this later. We always work with such a version.

Example 2.1 *(Sample mean)*: Let $f(x, \theta) = (x - \theta)^2 - x^2$. Clearly $Q(\theta) = \theta^2 - 2\mathbb{E}(X)\theta$ which is minimized uniquely at $\theta_0 = \mathbb{E}(X)$. Its unique M-estimator is the sample mean. □

Example 2.2 *(U-statistics as M_m-estimators)*: For a function $h(x_1, \ldots, x_m)$ which is symmetric in its arguments, let

$$f(x_1 \ldots, x_m, \theta) = \left[\theta - h(x_1, \ldots, x_m)\right]^2 - \left[h(x_1, \ldots, x_m)\right]^2. \tag{2.4}$$

Then $\theta_0 = \mathbb{E}\, h(Y_1, \ldots, Y_m)$ and θ_n is the U-statistic with kernel h. So all U-statistics are M_m-estimators. In particular, the sample variance is an M_2-estimator. □

Example 2.3 *(Median and other quantiles)*: Suppose Y is a random variable with distribution function $\mathbb{F}$. For any $0 < p < 1$, the pth *quantile* of $\mathbb{F}$ is

$$\mathbb{F}^{-1}(p) = \inf\{x : \mathbb{F}(x) \geq p\}. \tag{2.5}$$

To see this as an M-parameter, let

$$f(x, \theta) = \mid x - \theta \mid - \mid x \mid - (2p - 1)\theta. \tag{2.6}$$

Note that $|f(x, \theta)| \leq 2|\theta|$ and hence $Q(\theta) = \mathbb{E}f(Y, \theta)$ is finite for all $\theta \in \mathbb{R}$. It is easy to check that,

$$f(x, \theta) = \theta\left[2\mathcal{I}_{\{x \leq 0\}} - 1\right] + 2\int_0^\theta \left[\mathcal{I}_{\{x \leq s\}} - \mathcal{I}_{\{x \leq 0\}}\right]ds - (2p - 1)\theta. \tag{2.7}$$

Hence

$$Q(\theta) = 2\int_0^\theta \mathbb{F}(s)ds - 2p\theta \quad \text{for all} \quad \theta \in \mathbb{R}. \tag{2.8}$$

$Q(\theta)$ is minimized at $\mathbb{F}^{-1}(p)$. This minimizer of $Q(\theta)$ is unique if $\mathbb{F}$ is *strictly increasing* at $\mathbb{F}^{-1}(p)$. If $p = 1/2$, it is called the *population median*.

Suppose $\{Y_1, \ldots, Y_n\}$ is an i.i.d. sample from $\mathbb{F}$. Let $\mathbb{F}_n$ be the e.c.d.f. Then the *sample p-th quantile*, given by

$$\mathbb{F}_n^{-1}(p) = \inf\{x : \mathbb{F}_n(x) \geq p\}, \tag{2.9}$$

is a (measurable) minimiser of $Q_n(\theta)$. Unlike the sample mean, it is *not* necessarily unique. If $p = 1/2$, we get the sample median. □

Example 2.4 *(Hodges-Lehmann measure of location)*: Let $Y_1, \ldots, Y_n$ be i.i.d. observations from $\mathbb{F}$. Let $\mathbb{G}$ be the distribution of $\frac{Y_1 + Y_2}{2}$. Instead of the usual mean as a measure of location, consider the M_2-parameter $\theta_0 = \mathbb{G}^{-1}(1/2)$ (assume that $\mathbb{G}$ is strictly increasing at θ_0).

The *median* of $\{\frac{Y_i + Y_j}{2}, \quad 1 \leq i < j \leq n\}$, the *sample measure of location*, is an M_2-estimator of θ_0. Here $m = 2$ and

$$f(x_1, x_2, \theta) = \mid \frac{x_1 + x_2}{2} - \theta \mid - \mid \frac{x_1 + x_2}{2} \mid. \tag{2.10}$$

□

Example 2.5 *(Robust measure of scale/dispersion)*: Usually, M-estimators are thought of as measures of location. However, M_m-estimators encompass measures of both, location and scale. The variance as a measure of dispersion is influenced by extreme observations. To address this problem, Bickel and Lehmann (1979) considered the distribution of $|Y_1 - Y_2|$ and took its median to be a measure of dispersion. As in the previous examples, this is an M_2-parameter provided the distribution of $|Y_1 - Y_2|$ is strictly increasing at the median. Here $m = 2$ and

$$f(x_1, x_2, \theta) = \big|\, |x_1 - x_2| - \theta \,\big| - |x_1 - x_2|\,. \tag{2.11}$$

The corresponding estimate is the median of $\{|Y_i - Y_j|, 1 \le i < j \le n\}$ and as before, need not be unique.   □

Example 2.6 *(U-quantiles)*: The ideas of the previous two examples can be extended to define U-quantiles of Choudhury and Serfling (1988). Suppose that $h(x_1, \ldots, x_m)$ is a symmetric kernel. Define

$$f(x_1, \ldots, x_m, \theta) = \big|\, h(x_1, \ldots, x_m) - \theta \,\big| - |h(x_1, \ldots, x_m)|\,. \tag{2.12}$$

Then θ_0, the minimizer of $\mathbb{E}[f(Y_1, \cdots, Y_m, \theta)]$, is called a U-*median*. It is the unique minimiser if the distribution of $h(Y_1, \ldots, Y_m)$ is strictly increasing at θ_0. Its sample version is the median of $\{h(Y_{i_1}, Y_{i_2}, \ldots, Y_{i_m}), 1 \le i_1 < \cdots < i_m \le n\}$. Other U-*quantiles* can be defined in a way similar to the sample quantiles in Example 2.3. Note that just like the sample quantiles, the sample minimizers, in general, are not unique. Multivariate U-quantiles defined by Helmers and Hušková (1994) are also M_m-estimates.   □

Now suppose we have multivariate observations. Then there are several reasonable definitions of "median". The reader may consult the excellent article of Small (1990) for an introduction to the various notions of median/location for multivariate observations. We shall discuss some of these which fit into our framework.

Example 2.7 *(L₁-median)*: Suppose $\mathbf{Y} = (Y_1, \ldots, Y_d)$ is a $d \ge 2$ dimensional random vector with corresponding probability distribution $\mathbb{P}$. Let

$$f(x, \theta) = \Big[\sum_{k=1}^{d} (x_k - \theta_k)^2\Big]^{1/2} - \Big[\sum_{k=1}^{d} x_k^2\Big]^{1/2}\,. \tag{2.13}$$

Note that $Q(\theta) = \mathbb{E}f(\mathbf{Y}, \theta)$ is finite if $\mathbb{E}|\mathbf{Y}| < \infty$, where $|a|$ is the Euclidean norm of the vector a. It can be shown that if $\mathbb{P}$ does not put all its mass on a hyperplane (that is, if $\mathbb{P}\left[\sum_{i=1}^{d} C_i Y_i = C\right] \neq 1$ for any choice of real numbers $(C, C_1, \ldots C_d)$), then $Q(\theta)$ is minimized at a *unique* θ_0 (see Kemperman (1987)). This θ_0 is called the L_1-median. The corresponding M-estimator is called the sample L_1-median. It is unique if $\{Y_1, \ldots, Y_d\}$ do not lie on a lower dimensional hyperplane. If $d = 1$, the L_1-median reduces to the usual median discussed in Example 2.3.

Later in Chapter 5 we illustrate the L_1-median with a real data application. The R package for this book, called UStatBookABSC also contains a function to obtain the L_1-median on general datasets. $\qquad\square$

Example 2.8 *(Oja-median)*: This multivariate median was introduced by Oja (1983). For any $(d + 1)$ points $x_1, x_2, \ldots, x_{d+1} \in \mathbb{R}^d$ for $d \geq 2$, the *simplex* formed by them is the smallest convex set containing these points. Let $\Delta(x_1, \ldots, x_d, x_{d+1})$ denote the absolute volume of this simplex. Let

$$f(x_1, \ldots, x_d, \theta) = \Delta(x_1, \ldots, x_d, \theta) - \Delta(x_1, \ldots, x_d, 0). \qquad (2.14)$$

Suppose $(Y_1, \ldots, Y_d)^T$ are d-dimensional i.i.d. random vectors with distribution function $\mathbb{F}$. If $\mathbb{E}|Y_1| < \infty$, then $Q(\theta) = \mathbb{E}\, f(Y_1, \ldots, Y_d, \theta)$ exists. It is known that if $\mathbb{F}$ does not have all its mass on a lower dimensional hyperplane, then there is a unique minimiser θ_0 of $Q(\theta)$ (see León and Massé (1993)). It is called the *Oja-median*. If $Y_1, \ldots, Y_n$ is an i.i.d. sample from $\mathbb{F}$, then any corresponding minimizer of $Q_n(\theta)$ is the sample Oja-median. It is unique if $Y_1, \ldots, Y_n$ do not lie on a lower dimensional hyperplane. $\qquad\square$

2.2 Convexity

Many researchers have studied the asymptotic properties of M-estimators and M_m-estimators. Early works on the asymptotic properties of M_1-estimators and M_2-estimators are Huber (1967) and Maritz et al. (1977). Using conditions similar to Huber (1967), Oja (1984) proved the *consistency and asymptotic normality* of M_m-estimators. His results apply to some of the estimators that we have presented above.

We emphasize that all examples of f we have considered so far have a common feature. They are all *convex* functions of θ. Statisticians prefer to work with convex loss functions for various reasons. We shall make this blanket assumption here. This does entail some loss of generality. But convexity leads to a significant simplification in the study of M_m-estimators while at the same time, still encompassing a huge class of estimators.

Several works have assumed and exploited this convexity in similar contexts. Perhaps the earliest use of this convexity was by Heiler and Willers (1988) in linear regression models. See also Hjort and Pollard (1993). For $m = 1$, Haberman (1989) established the consistency and asymptotic normality of θ_n and Niemiro (1992) established a Bahadur type representation (linearization) $\theta_n = \theta_0 + S_n/n + R_n$ where R_n is of suitable order *almost surely* and S_n is the partial sum of a sequence of i.i.d. random variables. This was extended by Bose (1998) to M_m-estimators. In the next sub-sections, we shall exploit the convexity heavily and establish some large sample properties of M_m-estimates.

Even though our setup covers a lot of interesting multivariate location and scale estimators, we must emphasize that it does *not* cover several other important estimators such as the medians of Liu (1990), Tukey (1975), Rousseeuw (1985) and others, since the convexity condition is not satisfied. More general approaches in the absence of convexity are provided by Jurečková (1977), Jurecková and Sen (1996) and Chatterjee and Bose (2005). See also de la Peña and Giné (1999) page 279.

Now note that since f is convex in θ, it has a *sub-gradient* $g(x, \theta)$ (see Rockafellar (1970), page 218). This sub-gradient has the property that for all α, β, x,

$$f(x, \alpha) + (\beta - \alpha)^T g(x, \alpha) \le f(x, \beta). \tag{2.15}$$

It can be checked that a vector $\gamma \in \mathbb{R}^d$ is a sub-gradient of $f(z, \cdot)$ at α if and only if $h(z, \gamma) \ge 0$ (see Rockafellar (1970), page 214), where

$$h(z, \gamma) = \inf_{\beta} \left[f(z, \beta) - f(z, \alpha) - (\beta - \alpha)^T \gamma \right]. \tag{2.16}$$

In the next sub-section we will show that we can always choose a measurable version of the sub-gradient. If f is differentiable, then this sub-gradient is simply the ordinary derivative. This sub-gradient will be crucial to us.

Example 2.9: (i) For the usual median, it can be checked that a sub-gradient is given by

$$g(x, \theta) = \begin{cases} 1 & \text{if} \quad \theta > x \\ 0 & \text{if} \quad \theta = x \\ -1 & \text{if} \quad \theta < x. \end{cases} \tag{2.17}$$

(ii) For the L_1-median, it can be checked that a sub-gradient is given by

$$g(x,\theta) = \begin{cases} \dfrac{\theta - x}{|\theta - x|} & \text{if } \theta \neq x \\ 0 & \text{if } \theta = x. \end{cases} \qquad (2.18)$$

$\square$

2.3 Measurability

As Examples 2.3 and 2.6 showed, an M_m-estimator is not necessarily unique. However, it can be shown by using the convexity assumption, that a *measurable minimizer* can always be chosen. This can be done by the following selection theorem and its corollary. The asymptotic results that we will discuss later, hold for *any* measurable sequence of minimizers of $\{Q_n(\theta)\}$.

At the heart of choosing a sequence of measurable minimizers is the idea of *measurable selections*. This is a very interesting topic in mathematics and there are many selection theorems in the literature. See for example Castaing and Valadier (1977). Γ is said to be a *multifunction* if it assigns a subset $\Gamma(z)$ of $\mathbb{R}^d$ to each z. A function $\sigma : \mathbf{Z} \to \mathbb{R}^d$ is said to be a *selection* of Γ if $\sigma(z) \in \Gamma(z)$ for every z. If $\mathbf{Z}$ is a measurable space, then σ is said to be a *measurable selection* if $z \to \sigma(z)$ is a measurable function.

We quote the following theorem from Castaing and Valadier (1977). For its proof, see Theorem 3.6 and Proposition 3.11 in Section 3.2 there.

Theorem 2.1 (Castaing and Valadier (1977)). *(Selection theorem) Let Γ be a multifunction from a measurable space $\mathbf{Z}$ to a closed non-empty subset of $\mathbb{R}^d$. Let $\emptyset$ denote the empty set. If for each compact set K in $\mathbb{R}^d$, $\{z : \Gamma(z) \cap K \neq \emptyset\}$ is measurable, then Γ admits a measurable selection.*

Corollary 2.1. *Let $\mathbf{Z}$ be a measurable space and $q : \mathbf{Z} \times \mathbb{R}^d \to \mathbb{R}$ be a function. Assume $q(z, \cdot)$ is continuous for every z and $q(\cdot, \alpha)$ is measurable for every α. Then there is a measurable function $a : \mathbf{Z} \to \mathbb{R}^d$ such that*

$$q(z, a(z)) = \inf_{\alpha} q(z, \alpha),$$

whenever the inf *is in the range of $q(z, \cdot)$, otherwise $a(z)$ is taken to be some fixed number.*

Proof of Corollary 2.1: To begin with, note that

$$\inf_{\alpha \in A} q(z, \alpha) : \mathbf{Z} \to \mathbb{R} \cup \{-\infty\} \qquad (2.19)$$

for any subset A of $\mathbb{R}^d$. Indeed, $\inf_{\alpha \in A}$ can be replaced by $\inf_{\alpha \in C}$, where C is a *countable dense subset* of A, because $q(z, \cdot)$ is continuous. Let

$$\Gamma(z) = \{\beta : q(z, \beta) = \inf_{\alpha} q(z, \alpha)\}. \qquad (2.20)$$

We have

$$\Gamma(z) \neq \emptyset \Leftrightarrow \inf_{\alpha} q(z, \alpha) = \inf_{|\alpha| \leq n} q(z, \alpha) \quad \text{for some} \quad n. \qquad (2.21)$$

This is because the right side infimum is certainly in the range of $q(z, \cdot)$. Thus,

$$\mathbf{Z}_0 = \{z : \Gamma(z) \neq \emptyset\} \qquad (2.22)$$

is a measurable set. Define

$$a(z) = \quad \text{some constant, for all} \quad z \notin \mathbf{Z}_0 \qquad (2.23)$$

and consider Γ on $\mathbf{Z}_0$. Since $\Gamma(z)$ is always a closed subset of $\mathbb{R}^d$, it is enough to observe that for each compact $K, \{z : \Gamma(z) \bigcap K \neq \emptyset\}$ is equal to $\{z : \inf_{\alpha} q(z, \alpha) = \inf_{\alpha \in K} q(z, \alpha)\}$ and, consequently, it is a measurable set.

The existence of a measurable selection $a : \mathbf{Z}_0 \to \mathbb{R}^d$ of Γ follows now from Theorem 2.1. $\qquad \square$

We now show how we can apply the above corollary to obtain a measurable minimiser θ_n in (2.3). Suppose $f(x_1, \ldots, x_m, \theta)$ is a function on $\mathcal{Y}^m \times \mathbb{R}^d$ which is measurable in $(x_1, \ldots, x_m)$ and convex in θ. Note that convexity automatically implies continuity in θ.

Suppose $\{Y_1, \ldots, Y_n\}, n \geq m$ are i.i.d. $\mathcal{Y}$ valued random variables. On $\mathcal{Y}^n$ consider the function $q(\cdot, \alpha) = Q_n(\alpha)$ and apply Corollary 2.1, to get a random vector $\alpha_n = a(Y_1, \ldots, Y_n)$ that satisfies (2.3). Take θ_n to be equal to this α_n. Hence note that θ_n is measurable.

Now, having proved that there is at least one measurable θ_n, henceforth, we always work with a measurable θ_n. The asymptotic properties of θ_n are intimately tied to the sub-gradient g of f. So we need to make sure the existence of a measurable sub-gradient. This follows immediately from the following Corollary.

Corollary 2.2. *Let $\mathbf{Z}$ be a measurable space and $f : \mathbf{Z} \times \mathbb{R}^d \longrightarrow \mathbb{R}$. Assume that $f(z, \cdot)$ is convex for every z, and $f(\cdot, \alpha)$ is measurable for every α. Then there is $g : \mathbf{Z} \times \mathbb{R}^d \longrightarrow \mathbb{R}^d$ such that $g(z, \cdot)$ is a sub-gradient of $f(z, \cdot)$ for every z and $g(\cdot, \alpha)$ is measurable for every α.*

Proof of Corollary 2.2: Fix α. Recall the criterion presented in (2.16) for any vector to be a sub-gradient. Now, for every $z, h(z, \cdot)$ is a concave and finite-valued function; hence it is continuous. For every γ, $h(\cdot, \gamma)$ is measurable, because the infimum can be taken over β in a countable dense set, as in the preceding proof (see the arguments just before (2.20)).

Denoted by $\Gamma(z)$ the set of all *sub-derivatives* (i.e., the set of all sub-gradients) of $f(z, \cdot)$ at α. Then $\Gamma(z)$ is a non-empty closed set. Now note that for each compact $K, \{z : \Gamma(z) \cap K \neq \emptyset\}$ and is measurable, for it is equal to $\{z : \sup_{\gamma \in K} h(\gamma, z) \geq 0\}$. Now we can apply the Selection Theorem 2.1 to complete the proof. $\qquad\square$

2.4 Strong consistency

Definition 2.3: A sequence of estimators $\{\theta_n\}$ of θ_0 is said to be *strongly consistent* if $\theta_n \xrightarrow{a.s.} \theta_0$ as $n \to \infty$.

Interestingly, all that is needed to guarantee strong consistency of an M_m-estimator are the following minimal assumptions. They will be in force throughout this chapter.

    (I) $f(x_1, \ldots, x_m, \theta)$ is measurable in $(x_1, \ldots, x_m)$ and convex in θ.

    (II) $Q(\theta)$ is finite for all θ.

    (III) θ_0 exists and is unique.

Incidentally the parameter space need not always be the entire $\mathbb{R}^d$ and may be restricted to an appropriate convex subset. If (I) and (II) are satisfied on such a subset, then all the results we give below remain valid if θ_0 is an interior point of this subset.

Theorem 2.2. *(Strong consistency) Under Assumptions I, II and III, for* *any sequence of measurable minimizers $\{\theta_n\}$ of $Q_n(\theta)$,*

$$\theta_n \xrightarrow{a.s.} \theta_0 \quad as \quad n \to \infty. \tag{2.24}$$

This Theorem in particular implies that *all* the estimators introduced so far in our examples are strongly consistent as soon as we make sure that the minimal assumptions (II)–(III) hold.

To prove the Theorem, we need the following Lemma.

Lemma 2.1. *Suppose $\{h_n\}$ is a sequence of random convex functions on $\mathbb{R}^d$ which converges to a function h pointwise either almost surely or in probability. Then this convergence is uniform on any compact set of $\mathbb{R}^d$, respectively (a) almost surely, (b) in probability.*

Proof of Lemma 2.1: Recall that convex functions converge pointwise everywhere if they converge pointwise on a dense set. Moreover the everywhere convergence is uniform over compact sets. See Rockafellar (1970), Theorem 10.8 for additional details.

Let C be a countable dense set. To prove (a), it is just enough to observe that with probability 1, convergence $h_n(\alpha) \to h(\alpha)$ takes place for all $\alpha \in C$ and then apply the above criterion for convergence of convex functions.

To prove (b), consider an arbitrary sub-sequence of the sequence. For any fixed $\alpha \in C$, we can select a further sub-sequence, along which $h_n(\alpha) \to h(\alpha)$ holds almost surely. Now we can apply the Cantor diagonal method to get hold of one single sub-sequence $\{h_n\}$ which converges pointwise almost surely on C. Now apply (a) to conclude that this sub-sequence converges almost everywhere uniformly on compact sets. Since for *any* sub-sequence, we have exhibited a further sub-sequence which converges uniformly, almost surely on compact sets, the original sequence converges in probability uniformly on compact sets. This completes the proof.   □

Proof of Theorem 2.2: Note that by the SLLN for U-statistics, $Q_n(\alpha)$ converges to $Q(\alpha)$ for each α almost surely. By Lemma 2.1, this convergence is uniform on any compact set almost surely.

Let B be a ball of arbitrary radius ϵ around θ_0. If θ_n is not consistent, then there is an $\epsilon > 0$ and a set S in the probability space such that $\mathbb{P}(S) > 0$ and for each sample point in S, there is a sub-sequence of θ_n that lies outside this ball. We assume without loss that for each point in this set, the convergence of Q_n to Q also holds. For a fixed sample point, we continue to denote such a sequence by $\{n\}$.

Let θ_n^* be the point of intersection of the line joining θ_0 and θ_n with the ball B. Then for some sequence $0 < \gamma_n < 1$,

$$\theta_n^* = \gamma_n \theta_0 + (1 - \gamma_n)\theta_n.$$

By convexity of Q_n and the fact that θ_n is a minimizer of Q_n,

$$\begin{aligned} Q_n(\theta_n^*) &\leq \gamma_n Q_n(\theta_0) + (1 - \gamma_n)Q_n(\theta_n) \\ &\leq \gamma_n Q_n(\theta_0) + (1 - \gamma_n)Q_n(\theta_0) \\ &= Q_n(\theta_0). \end{aligned}$$

First note that the right side converges to $Q(\theta_0)$. Now, every θ_n^* lies on the compact set $\{\theta : |\theta - \theta_0| = \epsilon\}$. Hence there is a sub-sequence of $\{\theta_n^*\}$ which converges to, say θ_1. Since the convergence of Q_n to Q is uniform on compact sets, the left side of the above equation converges to $Q(\theta_1)$. Hence, $Q(\theta_1) \leq Q(\theta_0)$. This is a contradiction to the uniqueness of θ_0 since $|\theta_0 - \theta_1| = \epsilon$. This proves the Theorem. $\qquad\square$

2.5 Weak representation, asymptotic normality

We now give an *in probability* representation (linearization) result for M_m-estimators. This representation implies the asymptotic normality of M_m-estimators.

Let $g(x, \theta)$ be a measurable sub-gradient of f. It is easy to see by using (2.16), that under Assumption (II), the expectation of g is finite. Moreover, the gradient vector $\nabla Q(\theta)$ of Q at θ exists and

$$\nabla Q(\theta) = \mathbb{E}[g(Y_1, \ldots, Y_m, \theta)] < \infty. \tag{2.25}$$

Denote the matrix of second derivatives of Q at θ, whenever it exists, by $\nabla^2 Q(\theta)$. So

$$\nabla^2 Q(\theta) = \left(\left(\frac{\partial^2 Q(\theta)}{\partial \theta_i \partial \theta_j}\right)\right).$$

Let

$$H = \nabla^2 Q(\theta_0) \text{ and} \tag{2.26}$$

$$U_n = \binom{n}{m}^{-1} \sum_{1 \leq i_1 \cdots < i_m \leq n} g(Y_{i_1}, \ldots, Y_{i_m}, \theta_0). \tag{2.27}$$

Let N be an appropriate neighborhood of θ_0. We list two additional assumptions to derive a weak representation and asymptotic normality.

   (IV) $\mathbb{E}|g(Y_1, \ldots, Y_m, \theta)|^2 < \infty \ \forall \ \theta \in N$.
   (V) $H = \nabla^2 Q(\theta_0)$ exists and is positive definite.

The following Theorem is a consequence of the works of Haberman (1989) and Niemiro (1992) for $m = 1$, and Bose (1998) for general m.

Theorem 2.3. *Suppose Assumptions (I)–(V) hold. Then for any sequence of measurable minimizers $\{\theta_n\}$,*

(a) $\theta_n - \theta_0 = -H^{-1}U_n + o_\mathbb{P}(n^{-1/2})$
(b) $n^{1/2}(\theta_n - \theta_0) \xrightarrow{D} N(0, m^2 H^{-1} K H^{-1})$ *where*

$$K = \mathbb{V}\left(\mathbb{E}\left(g(Y_1, \ldots Y_m, \theta_0) | Y_2, \ldots, Y_m \right) \right)$$

is the dispersion matrix of the first projection of

$$g(Y_1, \ldots, Y_m, \theta_0).$$

All M_m-estimators given in Section 2.1 satisfy the conditions of Theorem 2.3 once we make reasonable assumptions on the underlying distribution. Hence the asymptotic normality of a huge collection of estimators follows. After we give the proof of the Theorem, we will illustrate its use through a discussion of the appropriate assumptions required in some specific cases.

Proof of Theorem 2.3: Using (2.16), for all α, β,

$$f(x, \alpha) + (\beta - \alpha)^T g(x, \alpha) \leq f(x, \beta), \tag{2.28}$$

$$f(x, \beta) + (\alpha - \beta)^T g(x, \beta) \leq f(x, \alpha). \tag{2.29}$$

Hence,

$$(\beta - \alpha)^T g(x, \alpha) \le f(x, \beta) - f(x, \alpha) \le (\beta - \alpha)^T g(x, \beta) \qquad (2.30)$$

or

$$0 \le f(x, \beta) - f(x, \alpha) - (\beta - \alpha)^T g(x, \alpha)$$
$$\le (\beta - \alpha)^T [g(x, \beta) - g(x, \alpha)]. \qquad (2.31)$$

Notice that $Q(\theta) = \mathbb{E}f(Y_1, Y_2, \ldots, Y_m, \theta)$ is finite. Consequently, it follows from (2.30) that $\mathbb{E}g(Y_1, Y_2, \ldots, Y_m, \theta)$ is finite for all θ. Moreover, based on (2.31), note that $\mathbb{E}g(Y_1, Y_2, \ldots, Y_m, \theta)$ serves as a subgradient of $Q(\theta)$. Now, when $Q(\theta)$ is differentiable, it follows that

$$\nabla Q(\theta) = \mathbb{E}g(Y_1, Y_2, \ldots, Y_m, \theta).$$

For the proof of this Theorem as well for those Theorems given later, assume without loss that $\theta_0 = 0$ and $Q(\theta_0) = 0$. As a consequence,

$$\nabla Q(0) = \mathbb{E}g(Y_1, Y_2, \ldots, Y_m, 0).$$

Let $S = \{s = (i_1, i_2, \ldots, i_m) : 1 \le i_1 < i_2 \cdots < i_m \le n\}$. For any $s \in S$, let $Y_s = (Y_{i_1}, \ldots, Y_{i_m})$ and

$$Y_{n,s} = f(Y_s, n^{-1/2}\alpha) - f(Y_s, 0) - n^{-1/2}\alpha^T g(Y_s, 0). \qquad (2.32)$$

Note that $V_n = \binom{n}{m}^{-1} \sum_{s \in S} Y_{n,s}$ is a U-statistic. From Exercise 25 of Chapter 1, using (2.31), it follows that

$$\mathbb{V}\left(\binom{n}{m}^{-1} \sum_{s \in S} Y_{n,s}\right) \le \frac{m}{n} \mathbb{E}\left[(Y_{n,s} - \mathbb{E}Y_{n,s})\right]^2$$
$$\le K \frac{m}{n} \mathbb{E}Y_{n,s}^2$$
$$\le K \frac{m}{n^2} \mathbb{E}\left[\alpha^t \{g(Y_{n,s}, n^{-1/2}\alpha) - g(Y_{n,s}, 0)\}\right]^2.$$

Let Z be identically distributed as any Y_s. Let

$$Z_n = \alpha^T \{g(Z, n^{-1/2}\alpha) - g(Z, 0)\}. \qquad (2.33)$$

Note that using (2.31), $Z_n \geq 0$.

Further,

$$
\begin{aligned}
&Z_{n+1} - Z_n \\
&= \alpha^T \Big[g(Z, (n+1)^{-1/2}\alpha) - g(Z,0) \Big] - \alpha^T \Big[g(Z, n^{-1/2}\alpha) - g(Z,0) \Big] \\
&= \alpha^T \Big[g(Z, (n+1)^{-1/2}\alpha) - g(Z, n^{-1/2}\alpha) \Big].
\end{aligned}
$$

However, by using (2.31),

$$
\Big[(n+1)^{-1/2}\alpha - n^{-1/2}\alpha) \Big]^T \Big[g(Z, (n+1)^{-1/2}\alpha) - g(Z, n^{-1/2}\alpha) \Big] \geq 0.
$$

Hence $Z_{n+1} - Z_n \leq 0$. That is, $\{Z_n\}$ is non-increasing.

Let $\lim Z_n = Z_0 \geq 0$. Thus $\mathbb{E}(Z_n) \downarrow \mathbb{E}(Z_0)$. Now,

$$
\begin{aligned}
\mathbb{E}Z_n &= \mathbb{E}\alpha^T \Big[g(Z, n^{-1/2}\alpha) - g(Z,0) \Big] \\
&= \mathbb{E}\alpha^T g(Z, n^{-1/2}\alpha) \\
&= \alpha^T \nabla Q(n^{-1/2}\alpha)
\end{aligned}
$$

and since the second partial derivatives exist (at 0),

$$
\nabla Q(n^{-1/2}\alpha) \to 0.
$$

Hence $\mathbb{E}Z_n \to 0$. Hence $Z_0 = 0$ a.s. and as a consequence, $\mathbb{E}Z_n^2 \to 0$. Noting that $\mathbb{E}\, Y_{n,s} = Q(n^{-1/2}\alpha)$, it follows that for each fixed α,

$$
\begin{aligned}
n\binom{n}{m}^{-1} &\sum_{s \in S}(Y_{n,s} - \mathbb{E}Y_{n,s}) \\
&= nQ_n\Big(\frac{\alpha}{\sqrt{n}}\Big) - nQ_n(0) - n^{1/2}\alpha^T U_n - nQ\Big(\frac{\alpha}{\sqrt{n}}\Big) \qquad (2.34) \\
&\xrightarrow{\mathbb{P}} 0. \qquad\qquad\qquad\qquad\qquad\qquad\qquad\qquad\qquad\qquad (2.35)
\end{aligned}
$$

On the other hand, by Assumption (V),

$$
nQ\big(\alpha/\sqrt{n}\big) \to \alpha^T H\alpha/2 \quad \text{for every } \alpha. \qquad (2.36)
$$

Now, due to convexity, by Lemma 2.1, the convergences in (2.35) and (2.36) are uniform on compact sets. Thus for every $\epsilon > 0$ and every $M > 0$, the inequality

$$\sup_{|\alpha| \leq M} |nQ_n(\alpha/\sqrt{n}) - nQ_n(0) - \alpha^T n^{1/2} U_n - \alpha^T H\alpha/2| < \epsilon \qquad (2.37)$$

holds with probability at least $(1 - \epsilon/2)$ for large n.
Define the quadratic form

$$B_n(\alpha) = \alpha^T n^{1/2} U_n + \alpha^T H\alpha/2. \qquad (2.38)$$

Its minimizer is $\alpha_n = -H^{-1} n^{1/2} U_n$ and by UCLT Theorem 1.1

$$\alpha_n \xrightarrow{\mathcal{D}} N(0, m^2 H^{-1} K H^{-1}). \qquad (2.39)$$

The minimum value of the quadratic form is

$$B_n(\alpha_n) = -n^{1/2} U_n^T H^{-1} n^{1/2} U_n^T /2. \qquad (2.40)$$

Further, $n^{1/2} U_n$ is bounded in probability. So we can select an M such that

$$\mathbb{P}\left[|-H^{-1} n^{1/2} U_n| < M - 1 \right] \geq 1 - \epsilon/2. \qquad (2.41)$$

The rest of the argument is on the intersection of the two events in (2.37) and (2.41), and that has probability at least $1 - \epsilon$.

Consider the convex function

$$A_n(\alpha) = nQ_n(\alpha/\sqrt{n}) - nQ_n(0). \qquad (2.42)$$

From (2.37),

$$A_n(\alpha_n) \leq \epsilon - n^{1/2} U_n^T H^{-1} n^{1/2} U_n^T /2 = \epsilon + B_n(\alpha_n). \qquad (2.43)$$

Now consider the value of A_n on the sphere $\{\alpha : |\alpha - \alpha_n| = T\epsilon^{1/2}\}$ where T will be chosen. Again by using (2.37), on this sphere, its value is *at least*

$$B_n(\alpha) - \epsilon. \qquad (2.44)$$

Comparing the two bounds in (2.43) and (2.44), and using the condition that α lies on the sphere, it can be shown that the bound in (2.44) is always strictly larger than the one in (2.43) once we choose $T = 4[\lambda_{min}(H)]^{-1/2}$

where λ_{min} denotes the minimum eigenvalue.

On the other hand A_n has the minimizer $n^{1/2}\theta_n$. So, using the fact that A_n is convex, it follows that its minimizer satisfies $|n^{1/2}\theta_n - \alpha_n| < T\epsilon^{1/2}$. Since this holds with probability at least $(1 - \epsilon)$ where ϵ is arbitrary, the first part of the theorem is proved. The second part now follows from the multivariate version of UCLT Theorem 1.1.   □

Example 2.10*(Maximum likelihood estimator)*: Under suitable conditions, the maximum likelihood estimator (m.l.e.) is weakly or strongly consistent and asymptotically normal. See for example van der Vaart and Wellner (1996) for sets of conditions under which this is true. If we are ready to assume that the log-likelihood function is concave in the parameter, then these claims follow from the above theorem.   □

Example 2.11*(Sample quantiles)*: Recall (2.8) from Example 2.3. If $\mathbb{F}$ is continuous at a point θ, then $Q'(\theta) = 2\mathbb{F}(\theta) - 2p$.

Further if $\mathbb{F}$ is differentiable at θ_0 with derivative $f(\theta_0)$, then

$$H = Q''(\theta_0) = 2\,f(\theta_0)\,(> 0 \quad \text{if} \quad f(\theta_0) > 0). \tag{2.45}$$

Additionally,

$$g(x,\theta) = \mathcal{I}_{\{\theta \geq x\}} - \mathcal{I}_{\{\theta \leq x\}} - (2p - 1) = 2\mathcal{I}_{\{\theta \geq x\}} - \mathcal{I}_{\{x=\theta\}} - 2p. \tag{2.46}$$

Since g is bounded, Assumption (IV) is trivially satisfied. Thus all the conditions (I)–(V) are satisfied.

Moreover

$$K = \mathbb{V}\left[2\mathcal{I}_{\{\theta_0 \geq Y_1\}}\right] = 4\,\mathbb{V}\left[\mathcal{I}_{\{Y_1 \leq \theta_0\}}\right] = 4p(1 - p). \tag{2.47}$$

Hence if $f(\theta_0) > 0$, then the sample p-th quantile $\theta_n = \mathbb{F}_n^{-1}(p)$ satisfies

$$n^{1/2}(\theta_n - \theta_0) \xrightarrow{\mathcal{D}} N\left(0,\; p(1 - p)(f^2(\theta_0))^{-1}\right). \tag{2.48}$$

□

Incidentally, if the assumptions of Theorem 2.3 are not satisfied, the limiting distribution of the M_m-estimate need not be normal. Smirnov (1949) (translated in Smirnov (1952)) had studied the sample quantiles in such *non-regular* situations in complete details, identifying the class of distributions possible. Jurečková (1983) considered general M-estimates in non-regular situations. See also Bose and Chatterjee (2001a).

Example 2.12 *(U-quantiles, generalized order statistic)*: As in Example 2.11, let $h(x_1, \ldots, x_m)$ be a symmetric kernel and let

$$f(x_1, \ldots, x_m, \theta) = |h(x_1, \ldots, x_m) - \theta| - |h(x_1, \ldots, x_m)| - (2p - 1)\theta.$$

Let $Y_1, \ldots, Y_n$ be i.i.d. with distribution $\mathbb{F}$. Let us use the notation $\mathbb{K}$ to denote the c.d.f. of $h(Y_1, \ldots, Y_m)$. Let $\theta_0 = \mathbb{K}^{-1}(p)$ be the (unique) p-th quantile of $\mathbb{K}$. If $\mathbb{K}$ is differentiable at θ_0 with a positive density $k(\theta_0)$, then Assumption (V) holds with $H = 2k(\theta_0)$. The gradient vector is given by

$$g(x, \theta) = 2\mathcal{I}_{\{\theta \geq h(x)\}} - \mathcal{I}_{\{\theta = h(x)\}} - 2p. \tag{2.49}$$

This is bounded and hence (IV) holds trivially.

Let

$$\mathbb{K}_n(y) = \binom{n}{m}^{-1} \sum_{1 \leq i_1 < \cdots < i_m \leq n} \mathcal{I}_{\{h(Y_{i_1}, \ldots, Y_{i_m}) \leq y\}} \tag{2.50}$$

be the *empirical distribution*. The M_m-estimate is then $\mathbb{K}_n^{-1}(p)$, the pth-quantile of $\mathbb{K}_n$.

By application of Theorem 2.3,

$$n^{1/2}(\mathbb{K}_n^{-1}(p) - \mathbb{K}^{-1}(p)) \xrightarrow{D} N\left(0, \left(p(1-p)(k(\mathbb{K}^{-1}(p)))\right)^{-1}\right). \tag{2.51}$$

Particular cases of this result are the following four estimates.

(i) *Univariate Hodges-Lehmann estimator* of Hodges and Lehmann (1963) where

$$h(Y_1, \ldots, Y_m) = m^{-1}(Y_1 + \cdots + Y_m). \tag{2.52}$$

(ii) *Dispersion estimator* of Bickel and Lehmann (1979) where

$$h(Y_i, Y_j) = |Y_i - Y_j|. \tag{2.53}$$

(iii) *Regression coefficient estimator* introduced by Theil (see Hollander and

Wolfe (1973) pages 205-206) where (X_i, Y_i) are bivariate i.i.d. random variables and

$$h((X_i, Y_i), (X_j, Y_j)) = (Y_i - Y_j)/(X_i - X_j). \qquad (2.54)$$

(iv) *The location estimate of Maritz et al. (1977)* can also be treated in this way. Let β be any fixed number between 0 and 1. Let

$$L(x_1, x_2, \theta) = |\beta x_1 + (1-\beta)x_2 - \theta| + |\beta x_2 + (1-\beta)x_1 - \theta|. \qquad (2.55)$$

The minimizer of $\mathbb{E}\left[L(Y_1, Y_2, \theta) - L(Y_1, Y_2, 0)\right]$ is a measure of location of Y_i (Maritz et al. (1977)) and its estimate is the median of $\beta Y_i + (1-\beta)Y_j$, $i \neq j$ ($\beta = 1/2$ yields the Hodges-Lehmann estimator of order 2). Conditions similar to above guarantee asymptotic normality for this estimator. □

Example 2.13 *(L_1-median, $d \geq 2$):* The L_1-median θ_0 was defined in Example 2.7. If the dimension $d = 1$, then the L_1-median coincides with the usual median whose asymptotic normality was discussed in Example 2.11. So we assume that $d \geq 2$. Recall that the gradient vector is

$$g(x, \alpha) = \begin{cases} \dfrac{\alpha - x}{|\alpha - x|} & \text{if} \quad \alpha \neq x \\ \\ 0 & \text{if} \quad \alpha = x. \end{cases} \qquad (2.56)$$

Thus the sub-gradient g is a bounded function and hence Assumption (IV) is satisfied. Moreover, g is differentiable, except when $x = \theta$.

Assume that $\mathbb{P}(Y_1 = \theta) = 0$ for all θ. Then the matrix of partial derivatives of g is given by

$$h(x, \theta) = \frac{1}{|\theta - x|}\left(I - \frac{(\theta - x)(\theta - x)^T}{|\theta - x|^2}\right), \quad x \neq \theta. \qquad (2.57)$$

Consider the inverse moment condition

$$\mathbb{E}[|Y_1 - \theta_0|^{-1}] < \infty. \qquad (2.58)$$

This implies

$$\mathbb{E}|h(Y_1, \theta_0)| < \infty. \qquad (2.59)$$

Recall that $\nabla Q(\theta) = \mathbb{E}[g(Y_1, \theta)]$. By simple algebra, for $|x| \leq |\theta|$,

$$|g(x, \theta) - g(x, 0)| \leq 2|\theta|/|x|. \qquad (2.60)$$

Similarly, for $|x| > |\theta|$,

$$|g(x, \theta) - g(x, 0) - h(x, 0)\theta| \leq 5\frac{|\theta|^2}{|x|^2} + \frac{|\theta|^3}{|x|^3}. \qquad (2.61)$$

Using these two inequalities, and the inverse moment condition (2.58), it is easy to check that, the matrix H exists and can be evaluated as

$$H = \mathbb{E}[h(Y_1, \theta_0)]. \qquad (2.62)$$

$\square$

Example 2.14 *(Oja-median)*: Recall the Oja median defined in Example 2.8. Recall that $\Delta(x_1, \ldots, x_d, \theta)$ is the absolute volume of the simplex formed by $\{x_1, \ldots, x_d, \theta\}$. Let Y denote the $d \times d$ random matrix whose j-th column is $Y_j = (Y_{1j}, \ldots, Y_{dj})^T$, $1 \leq j \leq d$. That is,

$$Y = \begin{pmatrix} Y_{11} & Y_{12} & \cdots & Y_{1d} \\ Y_{21} & Y_{22} & \cdots & Y_{2d} \\ \cdot & \cdot & \cdot & \cdot \\ Y_{d1} & Y_{d2} & \cdots & Y_{dd} \end{pmatrix}.$$

Let $Y(i)$ be the $d \times d$ matrix obtained from Y by deleting its i-th row and replacing it by a row of 1's at the end. That is,

$$Y(i) = \begin{pmatrix} Y_{11} & Y_{12} & \cdots & Y_{1d} \\ Y_{21} & Y_{22} & \cdots & Y_{2d} \\ \cdot & \cdot & \cdot & \cdot \\ Y_{i-1,1} & Y_{i-1,2} & \cdots & Y_{i-1,d} \\ Y_{i+1,1} & Y_{i+1,2} & \cdots & Y_{i+1,d} \\ \cdot & \cdot & \cdot & \cdot \\ Y_{d1} & Y_{d2} & \cdots & Y_{dd} \\ 1 & 1 & \cdots & 1 \end{pmatrix}.$$

Finally let $M(\theta)$ be the $(d+1) \times (d+1)$ matrix obtained by augmenting the column vector $\theta = (\theta_1, \ldots, \theta_d)^T$ and a $(d+1)$ row vector of 1's respectively to the first column and the last row of Y. That is,

$$M(\theta) = \begin{pmatrix} Y_{11} & Y_{12} & \cdots & Y_{1d} & \theta_1 \\ Y_{21} & Y_{22} & \cdots & Y_{2d} & \theta_2 \\ \cdot & \cdot & \cdot & \cdot \\ Y_{d1} & Y_{d2} & \cdots & Y_{dd} & \theta_d \\ 1 & 1 & \cdots & 1 & 1 \end{pmatrix}.$$

Let $det(M)$ denote the determinant of the matrix M. It is easily seen that

$$f(Y_1, \ldots, Y_d, \theta) = |det(M(\theta))| - |det(M(0))| = |\theta^T T - Z| - |Z| \quad (2.63)$$

where

$$T = (T_1, \ldots, T_d)^T \quad T_i = (-1)^{i+1} det(Y(i)) \quad \text{and} \quad Z = (-1)^d det(Y). \quad (2.64)$$

Hence Q is well defined if $\mathbb{E}\,|Y_1| < \infty$. Further, if $g = (g_1, \ldots, g_d)$, then

$$g_i = T_i \cdot \text{sign}(\theta^T T - Z), \quad 1 \le i \le d \quad (2.65)$$

and has common features of the gradients of the sample mean $(g(x) = x)$ as well as of U-quantiles $(g(x) = $ sign function$)$, see Examples 2.12 and 2.13.

Assume that $\mathbb{E}|Y_1|^2 < \infty$. This implies $\mathbb{E}|T|^2 < \infty$ which in turn implies $\mathbb{E}|g_i|^2 < \infty$ and thus Assumption (IV) is satisfied.

To guarantee Assumption (V), suppose $\mathbb{F}$ is the distribution function of Y_1. First assume that $\mathbb{F}$ is continuous. Note that by arguments similar to those given in Example 2.11,

$$Q(\theta) - Q(\theta_0) = 2\mathbb{E}\big[\theta^T T\,\mathcal{I}_{\{Z \le \theta^T T\}} - \theta_0^T T\,\mathcal{I}_{\{Z \le \theta_0^T T\}}\big]$$
$$+ 2\mathbb{E}\big[Z\,\mathcal{I}_{\{Z \le \theta^T T\}} - Z\,\mathcal{I}_{\{Z \le \theta_0^T T\}}\big].$$

It easily follows that the i-th element of the gradient vector of $Q(\theta)$ equals

$$Q_i(\theta) = 2\mathbb{E}\big[T_i\,\mathcal{I}_{\{Z \le \theta^T T\}}\big]. \quad (2.66)$$

If further, $\mathbb{F}$ has a density, it follows that the derivative of $Q_i(\theta)$ with respect to θ_j is given by

$$Q_{ij}(\theta) = 2\mathbb{E}\big[T_i T_j f_{Z|T}(\theta^T T)\big] \tag{2.67}$$

where $f_{Z|T}(\cdot)$ denotes the conditional density of Z given T. Thus

$$H = \Big(\big(Q_{ij}(\theta_0)\big)\Big). \tag{2.68}$$

Clearly then Assumption (V) will be satisfied if we assume that, the density of $\mathbb{F}$ exists and H defined above exists and is positive definite. This condition is satisfied by many common distributions. □

Example 2.15: The *pth-order Oja-median* for $1 < p < 2$ is defined by minimizing

$$Q(\theta) = \mathbb{E}\Big[\Delta^p(Y_1, \ldots, Y_d, \theta) - \Delta^p(Y_1, \ldots, Y_d, 0)\Big] \tag{2.69}$$

where Δ is as in Example 2.8. The quantities g_i and H are now given by

$$g_i(\theta) = pT_i|\theta^T T - Z|^{p-1}\text{sign}\,(\theta^t T - Z), \; i = 1, \ldots, d, \tag{2.70}$$

$$H = \Big(\big(h_{ij}\big)\Big) = p(p-1)\Big(\big(\mathbb{E}[T_i T_j \theta_0^T T - Z|^{p-2}]\big)\Big). \tag{2.71}$$

Now it is easy to formulate conditions for the asymptotic normality of the pth-order Oja-median. □

2.6   Rate of convergence

In the previous section, we have shown that the leading term of an M_m-estimator is a U-statistic. Recall that in Chapter 1, we have proved some rate of convergence results for U-statistics under additional conditions on the kernel. Hence we take a cue from those results to demonstrate that when the conditions of Section 2.5 are appropriately strengthened, then the results on consistency and asymptotic normality can be sharpened considerably.

Let N be a neighborhood of θ_0 and $r > 1$ and $0 \le s < 1$ be such that the following assumptions hold. Further restrictions on r and s as needed are given in the forthcoming theorems.

(VIa) $\mathbb{E}\Big[\exp\big(t|g(Y_1, \ldots, Y_m, \theta)|\big)\Big] < \infty \;\; \forall\, \theta \in N$ and some $t = t(\theta) > 0$.

(VIb) $\mathbb{E}|g(Y_1, \ldots, Y_m, \theta)|^r < \infty \ \forall \ \theta \in N$.

Theorem 2.4. *Suppose Assumptions (I)–(V) hold.*
(a) If Assumption (VIa) also holds, then for every $\delta > 0$, there exists an $\alpha > 0$ such that,

$$\mathbb{P}\left[\sup_{k \geq n} |\theta_k - \theta_0| > \delta\right] = O(\exp(-\alpha n)). \qquad (2.72)$$

(b) If Assumption (VIb) also holds with some $r > 1$, then for every $\delta > 0$,

$$\mathbb{P}\left[\sup_{k \geq n} |\theta_k - \theta_0| > \delta\right] = o(n^{1-r}) \text{ as } n \to \infty. \qquad (2.73)$$

Theorem 2.4(a) implies that the rate of convergence is exponentially fast. This in turn implies that $\theta_n \to \theta_0$ *completely*. That is, for every $\delta > 0$,

$$\sum_{n=1}^{\infty} \mathbb{P}\left[|\theta_n - \theta_0| > \delta\right] < \infty. \qquad (2.74)$$

Note that if $r < 2$, then Assumption (VIb) is *weaker* than Assumption (IV) needed for the asymptotic normality. If $r > 2$, then Assumption (VIb) is stronger than Assumption (IV) but weaker than Assumption (VIa), and still implies complete convergence.

Incidentally, the *last time* that the estimator is ϵ distance away from the parameter is of interest as ϵ approaches zero. See Bose and Chatterjee (2001b) and the references there for some information on this problem.

To prove Theorem 2.4, we need a Lemma, but first a definition.

Definition 2.4: (δ-triangulation.) Let A_0 and B be sets in $\mathbb{R}^d$. We say that B is a δ-*triangulation* of A_0 if every $\alpha \in A_0$ is a convex combination, $\sum \lambda_i \beta_i$ of points $\beta_i \in B$ such that $|\beta_i - \alpha| < \delta$ for all i.

Recall that a real valued function h is said to be a Lipschitz function with Lipschitz constant L if

$$|h(\alpha) - h(\beta)| \leq L|\alpha - \beta| \quad \text{for all} \quad \alpha, \beta.$$

Lemma 2.2. *Let $A \subset A_0$ be convex sets in $\mathbb{R}^d$ such that $|\alpha - \beta| > 2\delta$ whenever $\alpha \in A$ and $\beta \notin A_0$. Assume that B is a δ-triangulation of A_0. If* *h is a Lipschitz function on A_0 with Lipschitz constant L, and Q is a convex function on A_0 then*

$$\sup_{\beta \in B} |Q(\beta) - h(\beta)| < \epsilon \quad \text{implies} \quad \sup_{\alpha \in A} |Q(\alpha) - h(\alpha)| < 5\delta L + 3\epsilon. \tag{2.75}$$

Proof of Lemma 2.2: Consider an $\alpha \in A_0$ and write it as a convex combination $\sum \lambda_i \beta_i$ with $\beta_i \in B$ and $|\beta_i - \alpha| < \delta$. Since $Q(\beta_i) < h(\alpha) + \delta L + \varepsilon$, we have

$$Q(\alpha) \le \sum \lambda_i Q(\beta_i) < h(\alpha) + \delta L + \varepsilon. \tag{2.76}$$

On the other hand, to each $\alpha \in A$ there corresponds $\beta \in B$ such that $|\alpha - \beta| < \delta$ and thus $\alpha + 2(\beta - \alpha) = \gamma \in A_0$. From (2.76) it follows that

$$\begin{aligned}
Q(\alpha) &\ge 2Q(\beta) - Q(\gamma) > 2(h(\beta) - \varepsilon) - (h(\gamma) + \delta L + \varepsilon) \\
&> 2(h(\alpha) - \delta L - \varepsilon) - (h(\alpha) + 3\delta L + \varepsilon) \\
&= h(\alpha) - 5\delta L - 3\varepsilon.
\end{aligned} \tag{2.77}$$

The result now follows from (2.76) and (2.77). □

Proof of Theorem 2.4: We first prove part (b). Fix $\delta > 0$. Note that Q is convex and hence is continuous. Further, it is also Lipschitz, with Lipschitz constant L say, in a neighborhood of 0. Hence there exists an $\epsilon > 0$ such that $Q(\alpha) > 2\epsilon$ for all $|\alpha| = \delta$.

Fix α. By Assumption (VIb) and Theorem 1.5,

$$\mathbb{P}\left(\sup_{k \ge n} |Q_k(\alpha) - Q_k(0) - Q(\alpha)| > \epsilon\right) = o(n^{1-r}). \tag{2.78}$$

Now choose $\tilde{\epsilon}$ and $\tilde{\delta}$ both positive such that $5\tilde{\delta}L + 3\tilde{\epsilon} < \epsilon$. Let $A = \{\alpha : |\alpha| \le \delta\}$ and $A_0 = \{\alpha : |\alpha| \le \delta + 2\tilde{\delta}\}$. Let B be a *finite* δ-triangulation of A_0. Note that such a triangulation exists. From (2.78),

$$\mathbb{P}\left[\sup_{k \ge n} \sup_{\alpha \in B} |Q_k(\alpha) - Q_k(0) - Q(\alpha)| > \epsilon\right] = o(n^{1-r}). \tag{2.79}$$

Since $Q_k(\cdot)$ is convex, using Lemma 2.2 (with $h = Q_k$) and (2.79),

$$\mathbb{P}\left[\sup_{k \ge n} \sup_{|\alpha| \le \delta} |Q_k(\alpha) - Q_k(0) - Q(\alpha)| < 5\tilde{\delta}L + 3\tilde{\epsilon} < \epsilon\right] = 1 - o(n^{1-r}). \tag{2.80}$$

Suppose that the event in (2.80) occurs. Using the fact that $f_k(\alpha) = Q_k(\alpha) - Q_k(0)$ is convex, $f_k(0) = 0$, $f_k(\alpha) > \epsilon$ for all $|\alpha| = \delta$ and $Q(\alpha) > 2\epsilon$ for all $|\alpha| = \delta$, we conclude that $f_k(\alpha)$ attains its minimum on the set $|\alpha| \leq \delta$. This proves part (b) of the Theorem.

To prove part (a), we follow the argument given in the proof of part (b) but use Theorem 1.5(c) to obtain the required exponential rate. The rest of the proof remains unchanged. We omit the details. $\qquad\square$

Example 2.16 *(U-quantiles and L_1-median)*: Whenever the gradient is uniformly bounded, Assumption (VIa) is trivially satisfied. In particular, this is the case for U-quantiles (Example 2.6) and the L_1-median (Example 2.7), and then Theorem 2.4(a) is applicable. $\qquad\square$

Example 2.17 *(Oja-median)*: Recall the Oja-median defined in Example 2.8 and discussed further in Example 2.14. Note that the r-th moment of the gradient g is finite if the rth moment of T is finite which in turn is true if the rth moment of Y_1 is finite and then Theorem 2.4(b) is applicable. $\qquad\square$

2.7 Strong representation theorem

We now proceed to strengthen the asymptotic normality Theorem 2.3 by imposing further assumptions. As before, let N be an appropriate neighborhood of θ_0 while $r > 1$ and $0 \leq s < 1$ are numbers. Suppose that as $\theta \to \theta_0$,

(VII) $|\nabla Q(\theta) - \nabla^2 Q(\theta_0)(\theta - \theta_0)| = O(|\theta - \theta_0|^{(3+s)/2})$.

(VIII) $\mathbb{E}|g(Y_1, \ldots, Y_m, \theta) - g(Y_1, \ldots, Y_m, \theta_0)|^2 = O(|\theta - \theta_0|^{(1+s)})$.

(IX) $\mathbb{E}|g(Y_1, \ldots, Y_m, \theta)|^r = O(1)$.

Theorem 2.5. *Suppose Assumptions (I)–(V) and (VII)–(IX) hold for some $0 \leq s < 1$ and $r > (8 + d(1 + s))/(1 - s)$. Then almost surely as $n \to \infty$,*

$$n^{1/2}(\theta_n - \theta_0) = -H^{-1}n^{1/2}U_n + O(n^{-(1+s)/4}(\log n)^{1/2}(\log\log n)^{(1+s)/4}).$$

$$(2.81)$$

Theorem 2.5 holds for $s = 1$, with the interpretation that $r = \infty$ and g is bounded. This is of special interest and we state this separately.

Theorem 2.6. *Assume that g is bounded and Assumptions (I)–(V) and (VII)–(VIII) hold with s = 1. Then almost surely as $n \to \infty$,*

$$n^{1/2}(\theta_n - \theta_0) = -H^{-1}n^{1/2}U_n + O(n^{-1/2}(\log n)^{1/2}(\log\log n)^{1/2}). \quad (2.82)$$

The almost sure results obtained in Theorems 2.5 and 2.6 are by no means exact. We shall discuss this issue in some details later.

To prove the Theorems, we need a Lemma. It is a refinement of Lemma 2.2 on convex functions to the *gradient* of convex functions.

Lemma 2.3. *Let $A \subset A_0$ be convex sets in $\mathbb{R}^d$ such that $|\alpha - \beta| > 2\delta$ whenever $\alpha \in A$ and $\beta \notin A_0$. Assume that B is a δ-triangulation of A_0. Let k be an $\mathbb{R}^d$ valued Lipschitz function on A_0 with Lipschitz constant L. Let p be a sub-gradient of some convex function on A_0. Then*

$$\sup_{\beta \in B} |k(\beta) - p(\beta)| < \epsilon \text{ implies } \sup_{\alpha \in A} |k(\alpha) - p(\alpha)| < 4\delta L + 2\epsilon \quad (2.83)$$

Proof of Lemma 2.3: Assume $p(\alpha)$ is a sub-gradient of the convex function $h(\alpha)$. Let $e \in \mathbb{R}^d, |e| = 1$. For each $\alpha \in A$, the point $\alpha + \delta e$ can be written as a convex combination $\sum \lambda_i \beta_i$ with $\beta_i \in B$ and $|\beta_i - \alpha - \delta e| < \delta$. As a consequence, $|\beta_i - \alpha| < 2\delta$. Now, using the definition of a subgradient,

$$h(\alpha + \delta e) \leq \sum \lambda_i h(\beta_i) \leq \sum \lambda_i(h(\alpha) + (\beta_i - \alpha)^T p(\beta_i)). \quad (2.84)$$

Thus,

$$
\begin{aligned}
\delta e^T &p(\alpha) \\
&\leq h(\alpha + \delta e) - h(\alpha) \leq \sum \lambda_i(\beta_i - \alpha)^T p(\beta_i) \\
&\leq \sum \lambda_i \Big[(\beta_i - \alpha)^T k(\alpha) + |\beta_i - \alpha||k(\beta_i) - k(\alpha)| \\
&\quad + |\beta_i - \alpha||p(\beta_i) - k(\beta_i)| \Big] \\
&\leq \delta e^T k(\alpha) + (2\delta)^2 L + 2\delta\varepsilon.
\end{aligned}
$$

The lemma follows easily from the above inequality. $\qquad \square$

Proof of Theorem 2.5: Recall the notation S, s, Y_s introduced at the start of the proof of Theorem 2.3. Define

$$G(\alpha) = \nabla Q(\alpha), \ G_n(\alpha) = \binom{n}{m}^{-1} \sum_{s \in S} g(Y_s, \alpha) \qquad (2.85)$$

and

$$Y_{n,s} = g(Y_s, \frac{\alpha}{\sqrt{n}}) - g(Y_s, 0). \qquad (2.86)$$

Note that

$$\mathbb{E}(Y_{n,s}) = G(\frac{\alpha}{\sqrt{n}}), \qquad (2.87)$$

and

$$\binom{n}{m}^{-1} \sum_{s \in S} Y_{n,s} = G_n(\frac{\alpha}{\sqrt{n}}) - U_n. \qquad (2.88)$$

Let

$$l_n = (\log \log n)^{1/2}.$$

By (VIII), for any $M > 0$,

$$\sup_{|\alpha| \leq M l_n} \mathbb{E}|Y_{n,s}|^2 = O((n^{-1/2} l_n)^{1+s}). \qquad (2.89)$$

By using Theorem 1.6 with $v_n^2 = C^2 n^{-(1+s)/2} l_n^{1+s}$, for some K and D,

$$\sup_{|\alpha| \leq M l_n} \mathbb{P}\left[n^{1/2} |G_n(\frac{\alpha}{\sqrt{n}}) - U_n - G(\frac{\alpha}{\sqrt{n}})| > KC n^{-(1+s)/4} l_n^{(1+s)/2} (\log n)^{1/2} \right]$$

$$\leq D n^{1-r/2} C^{-r/2} n^{r(1+s)/4} l_n^{-r(1+s)/2} (\log n)^{r/2}$$

$$= D n^{1-r(1-s)/4} (\log n)^{r/2} (\log \log n)^{-r(1+s)/4}. \qquad (2.90)$$

This is the main probability inequality required to establish the Theorem. The rest of the proof is similar to the proof of Theorem 2.4. The refinements needed now are provided by the triangulation Lemma 2.3 and the LIL for U_n, Theorem 1.4.

Assumption (VII) implies that for each $M > 0$,

$$\sup_{|\alpha| \leq M l_n} \left| H\alpha - n^{1/2} G(\frac{\alpha}{\sqrt{n}}) \right| = O(n^{-(1+s)/4} (\log \log n)^{(3+s)/4}), \qquad (2.91)$$

and so the inequality (2.90), continues to hold when we replace $n^{1/2} G(\frac{\alpha}{\sqrt{n}})$

by $H\theta$ in the left side of (2.90).

Let

$$\epsilon_n = n^{-(1+s)/4}l_n^{(1+s)/2}(\log n)^{1/2}. \tag{2.92}$$

Consider a $\delta_n = n^{-(1+s)/4}(\log n)^{1/2}$ triangulation of the ball $B = \{\alpha : |\alpha| \leq Ml_n+1\}$. We can select such a triangulation consisting of $O(n^{d(1+s)/4})$ points. From the probability inequality above it follows that

$$|n^{1/2}G_n(\frac{\alpha}{\sqrt{n}}) - n^{1/2}U_n - H\alpha| \leq KC\epsilon_n \tag{2.93}$$

holds simultaneously for all α belonging to the triangulation with probability $1 - O(n^{d(1+s)/4+1-r(1-s)/4}(\log n)^{r/2})$. Now we use Lemma 2.3 to extend this inequality to all points α in the ball. Letting $K_1 = KC(2|H|+1)$, we obtain

$$\mathbb{P}\left[\sup_{|\alpha| \leq Ml_n} n^{1/2}|G_n(\frac{\alpha}{\sqrt{n}}) - U_n - n^{-1/2}H\alpha| > K_1\epsilon_n\right] \tag{2.94}$$

$$= O\left(n^{d(1+s)/4+1-r(1-s)/4}(\log n)^{r/2}\right). \tag{2.95}$$

Since $r > [8 + d(1+s)]/(1-s)$, the right side is summable and hence we can apply the Borel-Cantelli Lemma to conclude that almost surely, for large n,

$$\sup_{|\alpha| \leq Ml_n} |n^{1/2}G_n(\frac{\alpha}{\sqrt{n}}) - n^{1/2}U_n - H\alpha| \leq K_1\epsilon_n. \tag{2.96}$$

By the LIL for U-statistics given in Theorem 1.4, $n^{1/2}U_nl_n^{-1}$ is bounded almost surely as $n \to \infty$. Hence we can choose M so that

$$|n^{1/2}H^{-1}U_n| \leq Ml_n - 1$$

almost surely for large n. Now consider the convex function $nQ_n(n^{-1/2}\alpha) - nQ_n(0)$ on the sphere

$$S = \{\alpha : |\alpha + H^{-1}n^{1/2}U_n| = K_2\epsilon_n\}$$

where $K_2 = 2K_1[\inf_{|e|=1}e^T He]^{-1}$. Clearly using (2.96),

$$e^T n^{1/2}G_n(-H^{-1}n^{1/2}U_n + K_2\epsilon_n e) \geq e^T HeK_2\epsilon_n - K_1\epsilon_n \geq K_1\epsilon_n > 0,$$

and so the radial directional derivatives of the function are positive. This shows that the minimiser $n^{1/2}\theta_n$ of the function must lie within the sphere

$$|n^{1/2}\theta_n + H^{-1}n^{1/2}U_n| \leq K_2\epsilon_n \qquad (2.97)$$

with probability one for large n, proving Theorem 2.5. $\qquad\qquad\square$

Proof of Theorem 2.6: Let v_n and X_{ns} be as in the proof of Theorem 2.5. Let U_n be the U-statistic with kernel $X_{ns} - \mathbb{E}X_{ns}$ which is now bounded since g is bounded. By arguments similar to those given in the proof of Theorem 1.6 for the kernel h_{n1},

$$\mathbb{P}\Big[|n^{1/2}U_n| \geq v_n(\log n)^{1/2}\Big] \leq \exp\{-Kt(\log n)^{1/2} + t^2 n/k\}, \qquad (2.98)$$

provided $t \leq n^{-1/2}kv_n/2m_n$, where $k = [n/m]$ and m_n is bounded by C_0 say. Letting $t = K(\log n)^{1/2}$, it easily follows that the right side of the above inequality is bounded by $\exp(-Cn)$ for some c. The rest of the proof is same as the proof of Theorem 2.5. $\qquad\qquad\square$

Example 2.18 *(U-quantiles)*: These were defined in Example 2.6. Choudhury and Serfling (1988) proved a representation for them by using the approach of Bahadur (1966). Such a result now follows directly from Theorem 2.6. Notice that the sub-gradient vector given in Example 2.12 is bounded. Additionally, suppose that

(VIII)$'$ $\mathbb{K}$ has a density k which is continuous around θ_0.
It may then be easily checked that

$$\mathbb{E}|g(x,\theta) - g(x,\theta_0)|^2 \leq 4|\mathbb{K}(\theta) - \mathbb{K}(\theta_0)| = O(|\theta - \theta_0|). \qquad (2.99)$$

Thus (VIII) holds with $s = 0$.
It is also easily checked (see Example 2.12) that

$$\nabla Q(\theta) = \mathbb{E}g(Y_1,\theta) = 2\mathbb{K}(\theta) - 2p. \qquad (2.100)$$

Assume that
(VII)$'$ $\mathbb{K}(\theta) - \mathbb{K}(\theta_0) - (\theta - \theta_0)k(\theta_0) = O(|\theta - \theta_0|^{\frac{3}{2}})$ as $\theta \to \theta_0$.
    Then Assumption (VII) holds with $s = 0$.

It is easy to check that this implies $Q(\theta)$ is twice differentiable at $\theta = \theta_0$ with $H = \nabla^2 Q(\theta_0) = 2k(\theta_0)$. Thus, under Assumptions (VII)′ and (VIII)′, Theorem 2.6 holds for U-quantiles. The same arguments also show that the location measure of Maritz et al. (1977) also satisfies Theorem 2.6 under conditions similar to above.

Example 2.19(*Oja-median*): Recall the notation of Examples 2.8 and 2.14. The i-th element of the gradient vector of f is given by $g_i = T_i \cdot \text{sign}(\theta^T T - Z)$, $i = 1, \ldots, d$.

Assumption (VIII) is satisfied if

$$\mathbb{E}||Y||^2 \left[\mathcal{I}_{\{\theta^T T \leq Z \leq \theta_0^T T\}} + \mathcal{I}_{\{\theta_0^T T \leq Z \leq \theta^T T\}} \right] = O(|\theta - \theta_0|^{1+s}). \tag{2.101}$$

If $\mathbb{F}$ has a density then so does the conditional distribution of Z given T. By conditioning on T and from the experience of the univariate median, it is easy to see that Assumption (VIII) holds with $s = 0$ if this conditional density is bounded uniformly in $\theta^T T$ for θ in a neighborhood of θ_0 and $\mathbb{E}|T|^2 < \infty$. For the case $d = 1$, this is exactly Assumption (VIII)′ in Example 2.3.

To guarantee Assumption (VII), recall that if F has a density, derivative Q_{ij} of $Q_i(\theta)$ with respect to θ_j and the matrix H are given by

$$Q_{ij}(\theta) = 2\mathbb{E}\left[T_i T_j f_{Z|T}(\theta^T T) \right] \text{ and } H = \left(\left(Q_{ij}(\theta_0) \right) \right) \tag{2.102}$$

where $f_{Z|T}(\cdot)$ denotes the conditional density of Z given T. Hence Assumption (VII) will be satisfied if we assume that for each i, as $\theta \to \theta_0$,

$$\mathbb{E}\left[|Y_i \{ \mathbb{F}_{Z|T}(\theta^T T) - \mathbb{F}_{Z|T}(\theta_0^T T) - f_{Z|T}(\theta_0^T T)(\theta - \theta_0)^T) T \}| \right]$$
$$= O(|\theta - \theta_0|^{(3+s)/2}). \tag{2.103}$$

This condition is satisfied by many common densities. The other required Assumption (IX) is satisfied by direct moment conditions on T or Y.

By a similar approach, it is easy to formulate conditions under which Theorem 2.6 holds for pth-order Oja-median for $1 < p < 2$.

□

Example 2.20(*L_1-median, mth-order Hodges-Lehmann estimate and geometric quantiles in dimension $d \geq 2$*): Suppose $Y, Y_1, Y_2, \ldots, Y_n$ are i.i.d. d dimensional random variables.

(i) (L_1-median). Since results for the univariate median (and quantiles) are very well known (see for example Bahadur (1966), Kiefer (1967)), we confine our attention to the case $d \geq 2$.

Proposition 2.1. *Suppose θ_0 is unique. If for some $0 \leq s \leq 1$,*

$$\mathbb{E}[|Y_1 - \theta_0|^{-(3+s)/2}] < \infty, \tag{2.104}$$

then according as $s < 1$ or $s = 1$, the representation of Theorem 2.5 or 2.6 holds for the L_1-median with $U_n = n^{-1} \sum_{i=1}^{d} (Y_i - \theta_0)/|Y_i - \theta_0|$ and H defined in Example 2.13 earlier.

To establish the proposition, we verify the appropriate Assumptions. Recall the gradient vector given in Example 2.13 which is bounded. Hence Assumptions (I)–(V) and (IX) are trivially satisfied. Let $\mathbb{F}$ be the distribution of Y_1.

To verify Assumption (VIII), without loss of generality assume that $\theta_0 = 0$. Noting that g is bounded by 1 and $|g(x, \theta) - g(x, 0)| \leq 2|\theta|/|x|$, we have

$$\mathbb{E}|g(Y_1, \theta) - g(Y_1, 0)|^2 \leq 4|\theta|^2 \int_{|x|>|\theta|} |x|^{-2} d\mathbb{F}(x) + \int_{|x|<|\theta|} d\mathbb{F}(x)$$

$$\leq 4|\theta|^{1+s} \int_{|x|>|\theta|} |x|^{-(1+s)} d\mathbb{F}(x)$$

$$+ |\theta|^{1+s} \int_{|x|<|\theta|} |x|^{-(1+s)} d\mathbb{F}(x)$$

$$\leq 4|\theta|^{1+s} \mathbb{E}[|Y_1|^{-(1+s)}].$$

The moment assumption (2.104) assures that Assumption (VIII) is satisfied since $(1 + s) \leq (3 + s)/2$. Recall the function $h(x, \theta)$ and H defined in Example 2.13. Note that under our assumptions H is positive definite. By using arguments similar to those given in Example 2.13, it is easily seen that for $|x| \leq |\theta|$,

$$|g(x, \theta) - g(x, 0) - h(x, 0)\theta| \leq 4|\theta|/|x|. \tag{2.105}$$

Similarly, for $|x| > |\theta|$,

$$|g(x, \theta) - g(x, 0) - h(x, 0)\theta| \leq 6\frac{|\theta|^2}{|x|^2}. \tag{2.106}$$

Using these two inequalities, and taking expectation,

$$|\nabla Q(\theta) - \nabla Q(0) - H\theta| \le I_1 + I_2 \tag{2.107}$$

where

$$I_1 \le 4|\theta| \int_{|x| \le |\theta|} |x|^{-1} d\mathbb{F}(x) \le 2|\theta|^{(3+s)/2} \int_{|x| \le |\theta|} |x|^{-(3+s)/2} d\mathbb{F}(x) \tag{2.108}$$

and using the fact that $0 \le s \le 1$,

$$I_2 \le 6|\theta|^2 \int_{|x| > |\theta|} |x|^{-2} d\mathbb{F}(x) \le 6|\theta|^{(3+s)/2} \int_{|x| \ge |\theta|} |x|^{-(3+s)/2} d\mathbb{F}(x). \tag{2.109}$$

The inverse moment condition (2.104) assures that Assumption (VII) holds with $\nabla^2 Q(\theta_0) = H$. Thus we have verified all the conditions needed and the proposition is proved. $\qquad\square$

Let us investigate the nature of the inverse moment condition (2.104). If Y_1 has a density f bounded on every compact subset of $\mathbb{R}^d$ then $\mathbb{E}[|Y_1 - \theta|^{-2}] < \infty$ if $d \ge 3$ and $\mathbb{E}[|Y_1 - \theta_0|^{-(1+s)}] < \infty$ for any $0 \le s < 1$ if $d = 2$, and Theorem 2.5 is applicable. However, this boundedness or even the *existence* of a density as such is *not* needed if $d \ge 2$. This is in marked contrast with the situation for $d = 1$ where the existence of the density is required since it appears in the leading term of the representation. For most common distributions, the representation holds with $s = 1$ from dimension $d \ge 3$, and with some $s < 1$ for dimension $d = 2$. The weakest representation corresponds to $s = 0$ and gives a remainder $O(n^{-1/4}(\log n)^{1/2}(\log \log n)^{1/4})$ if $\mathbb{E}[|Y_1 - \theta|^{-3/2}] < \infty$.

The strongest representation corresponds to $s = 1$ and gives a remainder $O(n^{-1/2}(\log n)^{1/2}(\log \log n)^{1/2})$ if $\mathbb{E}[|Y_1 - \theta|^{-2}] < \infty$.

The moment condition (2.104) forces $\mathbb{F}$ to necessarily assign zero mass at the median. Curiously, if $\mathbb{F}$ assigns zero mass to an entire neighborhood of the median, then the moment condition is automatically satisfied.

Now assume that the L_1-median is zero and Y is dominated in a neighborhood of zero by a variable Z which has a radially symmetric density $f(|x|)$. Transforming to polar coordinates, the moment condition is satisfied if the integral of $g(r) = r^{-(3+s)/2+d-1}f(r)$ is finite. If $d = 2$ and f is bounded in a neighborhood of zero, then the integral is finite for all $s < 1$. If $f(r) = O(r^{-\beta})$, $(\beta > 0)$, then the integral is finite if $s < 2d - 3 - 2\beta$. In particular, if f is bounded $(\beta = 0)$, then any $s < 1$ is feasible for $d = 2$ and $s = 1$ for $d = 3$.

(ii) (Hodges-Lehmann estimate) The above arguments also show that if the moment condition is changed to $\mathbb{E}\big[|m^{-1}(Y_1 + \cdots + Y_m) - \theta_0|^{-(3+s)/2}\big] < \infty$, Proposition 2.1 holds for the Hodges-Lehmann estimator with

$$U_n = \binom{n}{m}^{-1} \sum_{1 \le i_1 < i_2 < \ldots < i_m \le n} g(m^{-1}(Y_{i_1} + \cdots + Y_{i_m}), \theta_0)). \qquad (2.110)$$

(iii) (Geometric quantiles) For any u such that $|u| < 1$, the u-th geometric quantile of Chaudhuri (1996) is defined by taking $f(\theta, x) = |x - \theta| - |x| - u^T\theta$. Note that $u = 0$ corresponds to the L_1-median. The arguments given in the proof of Proposition 2.1 remain valid and the representation of Theorem 2.5 or 2.6 hold for these estimates. One can also define the Hodges-Lehmann version of these quantiles and the representations would still hold. $\qquad \square$

2.7.1 Comments on the exact rate

The above results are quite strong even though there is scope for further improvement from the experience of the univariate case (see Kiefer (1967)). Unlike the case for $d = 1$, the *boundedness* or even the *existence* of a density as such is *not* needed if $d \ge 2$. Obtaining the exact order is a delicate and hard problem. The higher order asymptotic properties of the sample median for $d = 1$ was extensively studied with suitable conditions on the density by Bahadur (1966) and Kiefer (1967) via the fluctuations of the sample distribution function which puts mass n^{-1} at the sample values.

This approach has been used by several authors in other similar situations. For example, a representation for U-quantiles was proved by Choudhury and Serfling (1988) by studying the fluctuations of the distribution function which puts equal mass at all the $\binom{n}{m}$ points $h(Y_{i_1}, \ldots, Y_{i_m})$, $1 \le i_1 \le i_2 \le \ldots \le i_m \le n$. Chaudhuri (1992) proved a representation for the L_1-median and its Hodges-Lehmann version in higher dimensions by the same approach.

We note here that the theory of *empirical processes* serves as a very valuable tool in the study of properties of estimators of the type studied in this chapter. For instance, Arcones (1996) derived some *exact* almost sure rates for U-quantiles under certain "local variance conditions" by using empirical processes. See also Arcones et al. (1994). In particular the asymptotic normality of the Oja-median and the L_1-median were obtained by this approach. In a similar vein, Sherman (1994) derived maximal inequalities for degenerate

U-processes of order k, $k \geq 1$. These inequalities can be used to determine the limiting distribution of estimators that optimize criterion functions having U-process structure. However, we did not pursue empirical process arguments here, since that would require considerable additional technical developments.

Generally speaking, the exact rate depends on the nature of the function f. See Arcones and Mason (1997) for some refined almost sure results in general M-estimation problems. As an example, consider the L_1-median when $d = 2$. If the density of the observations exists in a neighborhood of the median, is continuous at the median and $\mathbb{E}\, g(Y_1, \theta)$ has a second order expansion at the median, then the *exact* almost sure order of the remainder is $O(n^{-1/2}(\log n)^{1/2}(\log \log n))$.

2.8 Exercises

1. Show that (2.8) is minimized at $\theta = \mathbb{F}^{-1}(p)$ and is unique if $\mathbb{F}$ is strictly increasing at $\mathbb{F}^{-1}(p)$. Find out all the minimizers if $\mathbb{F}$ is not strictly increasing at $\mathbb{F}^{-1}(p)$.

2. List the properties of the various medians given in Small (1990).

3. Convince yourself of the uniqueness claim for the L_1-median made in Example 2.7.

4. Convince yourself of the uniqueness claim for the Oja-median made in Example 2.8.

5. Show that under Assumption (II), the expectation of g is finite. Hint: use (2.15).

6. Show that under Assumption (II), the gradient vector $\nabla Q(\theta)$ of Q at θ exists and

$$\nabla Q(\theta) = \mathbb{E}[g(Y_1, \ldots, Y_m, \theta)] < \infty. \tag{2.111}$$

7. Argue, how, in the proof of Theorem 2.3, we can without loss of generality, assume $\theta_0 = 0$ and $Q(\theta_0) = 0$.

8. Refer to (2.43) and (2.44). Show that $B_n(\alpha_n) - \epsilon > A_n(\alpha_n)$ for all $\alpha \in \{\alpha : |\alpha - \alpha_n| = 2[\lambda_{min}(H)]^{-1/2}\epsilon^{1/2}\}$.

9. Verify that Theorem 2.3 holds for Example 2.12(iv).

10. For the L_1 median given in Example 2.9(ii), check that the gradient vector is indeed given by (2.56).

11. Show that (2.58) implies (2.59).

12. Verify that (2.62) holds.

13. Verify the calculations for the Oja median given in Example 2.14.

14. Formulate conditions for the asymptotic normality of the pth-order Oja-median.

15. Under suitable assumptions, state and prove an asymptotic bivariate normality theorem for the sample mean and the sample median.

Chapter 3

Introduction to resampling

3.1 Introduction

In the previous two chapters we have seen many examples of statistical parameters and their estimates. In general suppose there is a parameter of interest θ and observable data $\mathbf{Y} = (Y_1, \ldots, Y_n)$. The steps for statistical inference can be divided into three broad issues.

(I) How do we estimate θ from the data $\mathbf{Y}$?

(II) Given an estimator $\hat{\theta}_n$ of θ (that is, a function of $\mathbf{Y}$) how good is this estimator?

(III) How do we obtain confidence sets, test hypothesis and settle other such questions of inference about θ?

Question (I) on estimating θ from $\mathbf{Y}$ is fundamental to statistics and new ideas of estimation emerge as new problems appear across the statistical horizon. In the previous chapters we have dealt with this question for some classes of parameters.

The next two questions are equally important. A traditional way to answer (II) is to report the variance or the mean squared error of $\hat{\theta}_n$, which are functions of the distribution of $\hat{\theta}_n$. The answer to the last question is also typically based on the probability distribution of $\hat{\theta}_n$, that is, its sampling distribution.

Most often, computation of the exact sampling distribution of $\hat{\theta}_n$ or of the exact variance of $\hat{\theta}_n$ are intractable problems. Only in very limited number

© Springer Nature Singapore Pte Ltd. 2018 and Hindustan Book Agency 2018
A. Bose and S. Chatterjee, *U-Statistics, Mm-Estimators and Resampling*, Texts
and Readings in Mathematics 75, https://doi.org/10.1007/978-981-13-2248-8_3

of situations is it feasible to compute the exact sampling distribution or the exact variance of an estimator.

One such situation is where the data is n i.i.d. observations $Y_1, \ldots, Y_n$ from $N(\theta, \sigma^2)$, and the estimator for θ is the sample mean $\hat{\theta}_n = n^{-1} \sum_{i=1}^{n} Y_i$. If σ^2 is known, then $\hat{\theta}_n \sim N(\theta, n^{-1}\sigma^2)$. When σ^2 is unknown, an M_2-estimator, as seen in Chapter 2 is given by

$$\widehat{\sigma^2} = (n-1)^{-1} \sum_{i=1}^{n} (Y_i - \hat{\theta}_n)^2.$$

Using considerable ingenuity, W. S. Gossett, who wrote under the pen name *Student*, obtained the *exact* sampling distribution of $T_n = n^{1/2}(\hat{\theta}_n - \theta)/\hat{\sigma}$ (see Student (1908)). This distribution is now known as Student's t-distribution with $(n-1)$ degrees of freedom.

If the variables are not i.i.d. normal, the above result does not hold. More importantly, it is typically impossible to find a closed form formula for the distribution of T_n. While it is possible to obtain the sampling distribution of many other statistics, there is no general solution. In particular, the sampling distribution of the U-statistics and the M_m-estimates that we discussed in Chapters 1 and 2 are completely intractable in general.

Nevertheless, asymptotic solutions are available. It is often possible to suitably center and scale the estimator $\hat{\theta}_n$, which then converges in distribution. In Chapters 1 and 2 we have seen numerous instances of this where convergence happens to the normal distribution.

We have also seen in Chapter 2 how asymptotic normality was established by a *weak representation* result, so that the leading term of the centered statistic is a sum of i.i.d. variables and thus the usual CLT can be applied. This linearization was achieved by expending considerable technical effort. There still remains two noteworthy issues.

First, such a linearization may not be easily available for many estimates and the limit distribution need not be normal. Second, even if

$$a_n(\hat{\theta}_n - \theta) \xrightarrow{D} N(0, \mathbf{V})$$

for some $\mathbf{V} > 0$ and some sequence $a_n \to \infty$, the *asymptotic variance* $\mathbf{V}$ is unknown and may not even have a closed form expression. In many situations, estimation of $\mathbf{V}$ is very hard. Example 3.2 below on the use of the sample median as an estimator of the population median is one such case, since the asymptotic variance depends on the true probability density value at the unknown true population median.

This is where resampling comes in. It attempts to replace analytic derivations with the force of computations. We will introduce some of the popular resampling techniques and their properties in Section 3.5 below, but before that, in Section 3.2 we set the stage with three classical examples of problems where we may study statistical inference using the (i) finite sample exact distribution approach if available, (ii) asymptotics-driven approach, and (iii) resampling-based approach. Our discussion on the basic ideas of resampling are centered around these three examples. In Section 3.3 we define the notion of consistency of resampling plans in estimating the variance and the entire sampling distribution.

Then we introduce the quick and easy resampling technique, *jackknife*, which is aimed primarily at estimating the bias and variance of a statistic. The *bootstrap* is introduced in the context of estimating the sampling properties of the sample mean in Section 3.4.1. We also introduce the *Singh Property* which shows in what sense and how the bootstrap can produce better estimates than an asymptotics-based method. This is followed with a discussion on resampling for the sample median in Section 3.4.3. After some discussion on the principles and features of resampling in general in Section 3.3.2, we present in Section 3.5 several resampling methods that have been developed for use in linear regression. In Chapter 4 we will focus on resampling for U-statistics and M_m-estimates.

3.2   Three standard examples

In the last two chapters we have seen that a host of estimators may be linearized with a sample mean type leading term. The sample mean is the simplest smooth statistic. Our first example records the asymptotic normality of the mean and introduces the important idea of Studentization. Generally speaking, resampling techniques work best for smooth functions of the sample mean after an appropriate Studentization.

In Chapter 2 we dealt with M_m-estimators. One of the simplest non-smooth M-estimator is the sample median. This is our second benchmark example. Some resampling plans are not suited for such non-smooth situations.

Resampling techniques prove their worth more when we have non-i.i.d. data since calculation of properties of sampling distributions become more complicated as we move away from the i.i.d. structure. The simplest example of a non-i.i.d. model is the linear regression and that is our third benchmark example. We shall see later that there are many eminently reasonable resampling techniques available for such non-i.i.d. models.

Example 3.1 *(The mean)*: Suppose Y_i's are i.i.d. according to some distribution $\mathbb{F}$ on the real line with unknown mean θ and unknown variance σ^2; the parameter of interest is θ. Consider the estimator $\hat{\theta}_n = \sum_{i=1}^{n} Y_i/n$. Define $Z_n = n^{1/2}(\hat{\theta}_n - \theta)$, and let $\mathcal{L}_{Z_n}$ denote its probability distribution function. The CLT says that

$$\mathcal{L}_{Z_n} \xrightarrow{\mathcal{D}} N(0, \sigma^2). \tag{3.1}$$

If σ^2 is known, this *asymptotic Normal limit* distribution allows us to make (asymptotically valid) inferences about θ, for example, perform hypothesis tests on, or obtain a confidence interval for, θ. We will refer to statistics like Z_n that are centered, and scaled by constants like $n^{1/2}$ that do not depend on the data, as *normalized* $\hat{\theta}_n$.

If σ^2 is unknown, then an M_2-estimator for it is given by

$$\widehat{\sigma^2} = (n-1)^{-1} \sum_{i=1}^{n} \left(Y_i - \hat{\theta}_n\right)^2.$$

Now instead of the normalized statistic Z_n, we may use the *Studentized* statistic

$$T_n = n^{1/2}(\hat{\theta}_n - \theta)/\hat{\sigma}.$$

Let $\mathcal{L}_{T_n}$ denote the probability distribution function of T_n. If $\mathbb{F}$ is $N(\theta, \sigma^2)$, then $\mathcal{L}_{T_n}$ is the Student's t-distribution with $(n-1)$ degrees of freedom. In general, if $\mathbb{F}$ has finite positive variance, then

$$\mathcal{L}_{T_n} \xrightarrow{\mathcal{D}} N(0, 1).$$

This asymptotic distribution is free of parameters, and hence is obviously suitable for inference on θ.

The notable features here are (*i*) the centering, (*ii*) estimate of the asymptotic variance, (*iii*) Studentization and (*iv*) asymptotic normality. We shall see later that even in this basic case, appropriate resampling techniques can assure better accuracy than offered by the normal approximation. $\square$

Example 3.2 *(The median)*: Suppose the data is as in Example 3.1. However, the parameter of interest is now the *median* $\xi \in \mathbb{R}$. Recall that for any distribution $\mathbb{F}$, and for any $\alpha \in (0,1)$, the α-th quantile of $\mathbb{F}$ is defined as

$$\mathbb{F}^{-1}(\alpha) = \inf_{x \in \mathbb{R}} \{x \in \mathbb{R} : \mathbb{F}(x) \geq \alpha\}.$$

In particular, $\xi = \mathbb{F}^{-1}(1/2)$, is the *median* of $\mathbb{F}$.

Now suppose that $\mathbb{F}$ has a density f. Recall that $\mathbb{F}_n$ is the e.c.d.f. which assigns mass n^{-1} at each Y_i, $1 \leq i \leq n$. Let the sample median, $\hat{\xi}_n = \mathbb{F}_n^{-1}(0.5)$ be the estimator of ξ. We have seen in Chapter 2 that if $f(\xi) > 0$, then

$$n^{1/2}(\hat{\xi}_n - \xi) \longrightarrow N\left(0, \left(4f^2(\xi)\right)^{-1}\right). \tag{3.2}$$

In order to use this result, an estimate of $f(\xi)$ is required. Note that this is a non-trivial problem since the density f is unknown and that forces us to enter the realm of density estimation. This estimation also adds an extra error when using the asymptotic normal approximation (3.2) for inference.

We shall see later that when we use an appropriate resampling technique, this additional estimation step is completely avoided. $\square$

Example 3.3 *(Simple linear regression)*: Suppose the data is $\{(Y_i, x_i), i = 1, \ldots, n\}$, where $x_1, \ldots, x_n$ is a sequence of *known* constants. Consider the simple linear regression model

$$Y_i = \beta_1 + \beta_2 x_i + e_i. \tag{3.3}$$

We assume that the *error* or *noise terms* $e_1, \ldots, e_n$ are i.i.d. from some distribution $\mathbb{F}$, with $\mathbb{E}e_1 = 0$, and $\mathbb{V}e_1 = \sigma^2 < \infty$.

The random variable Y_i is the i-th *response*, while x_i is often called the i-th covariate. In the above, we considered the case where the x_i's are nonrandom, but random variables may also be used as covariates with minor differences in technical conditions that we discuss later. The above simple linear regression has one slope parameter β_2 and the intercept parameter β_1, and is a special case of the multiple linear regression, where several covariates with their own slope parameters may be considered.

It is convenient to express the multiple regression model in a linear algebraic notation. We establish some convenient notations first.

Suppose a^T denotes the transpose of the (vector or matrix) a, and let $\mathbf{Y} = (Y_1, \ldots, Y_n)^T \in \mathbb{R}^n$ denote the vector of responses, $X \in \mathbb{R}^n \times \mathbb{R}^p$ denote the matrix of covariates whose i-th row is the vector $\mathbf{x}_i = (x_{i1}, x_{i2}, \ldots, x_{ip})$, the regression coefficients are $\boldsymbol{\beta} = (\beta_1, \ldots, \beta_p)^T$, and the noise vector is $\mathbf{e} = (e_1, \ldots, e_n)^T \in \mathbb{R}^n$. In this notation, the multiple linear regression model is given by

$$\mathbf{Y} = X\boldsymbol{\beta} + \mathbf{e}. \tag{3.4}$$

The simple linear regression (3.3) can be seen as a special case of (3.4), with the choice of $p = 2$, $x_{i1} = 1$ and $x_{i2} = x_i$ for $i = 1, \ldots, n$.

If $\mathbb{F}$ is the Normal distribution, *i.e.*, if $e_1 \ldots, e_n$ are i.i.d. $N\left(0, \sigma^2\right)$, then we have the *Gauss-Markov model*. This is the most well-studied model for linear regression, and exact inference is tractable in this case. For example, if $\boldsymbol{\beta} = (\beta_1, \ldots, \beta_p)^T \in \mathbb{R}^p$ is the primary parameter of interest, it can be estimated by the *maximum likelihood* method using the normality assumption, and the sampling distribution of the resulting estimator can be described.

In the simple linear regression case, let

$$\bar{Y} = n^{-1} \sum_{i=1}^{n} Y_i, \quad \bar{x} = n^{-1} \sum_{i=1}^{n} x_i.$$

The *maximum likelihood estimate* (m.l.e.) of $\boldsymbol{\beta}$ is then given by

$$\hat{\beta}_1 = \bar{Y} - \hat{\beta}_2 \bar{x},$$
$$\hat{\beta}_2 = \frac{\sum_{i=1}^{n}(x_i - \bar{x})(Y_i - \bar{Y})}{\sum_{i=1}^{n}(x_i - \bar{x})^2}.$$

The m.l.e. for the multiple linear regression coefficient is given by

$$\hat{\boldsymbol{\beta}} = \left(X^T X\right)^{-1} X^T \mathbf{Y}. \tag{3.5}$$

The exact distribution of $\hat{\boldsymbol{\beta}}$ in the simple linear regression is given by

$$\hat{\boldsymbol{\beta}} \sim N\left(\begin{pmatrix} \beta_1 \\ \beta_2 \end{pmatrix}, \sigma^2 \left(n \sum_{i=1}^{n} x_i^2 - \left(\sum_{i=1}^{n} x_i\right)^2\right)^{-1} \begin{pmatrix} \dfrac{\sum_{i=1}^{n} x_i^2}{n} & -\sum_{i=1}^{n} x_i \\ -\sum_{i=1}^{n} x_i & n \end{pmatrix}\right).$$

For the multiple linear regression case, we have

$$\hat{\beta} \sim N\left(\beta, \sigma^2 (X^T X)^{-1}\right).$$

The above exact distribution may be used for inference when σ^2 is known. Even when it is not known, an exact distribution can be obtained, when we use the estimate of σ^2 given in (3.6).

The exact distribution depends on the assumption that $\mathbb{F} = N(0, \sigma^2)$. Even when this assumption is violated, the *principle of least squared errors* yields the same estimators $\hat{\beta}$ as above, although the above exact distribution is no longer valid. In this situation, we call $\hat{\beta}$ the *least squares estimate* (l.s.e.). Of course the l.s.e. may not be *optimal* when $\mathbb{F}$ is not $N(0, \sigma^2)$, but the issue of optimality under perfect distributional specification is moot when the distribution is unknown, as is the case in most practical situations.

The vector of *residuals* from the above multiple linear regression model fitting is defined as

$$\mathbf{r} = \mathbf{Y} - X\hat{\beta}.$$

Note that the residual vector is a statistic as it depends on $\mathbf{Y}$, X and $\hat{\beta}$ only, and is thus a computable quantity. In particular, it is *not* the same as the error e_i, which is an unknown, unobservable random variable. An estimate of the noise variance σ^2 is usually taken to be

$$\widehat{\sigma^2} = (n - p)^{-1} \sum_{i=1}^{n} r_i^2. \tag{3.6}$$

Though the exact distribution of $\hat{\beta}$ given in (3.5) is intractable when $\mathbb{F}$ is not Gaussian, we can establish asymptotic distribution results for $\hat{\beta}$. We describe only the case where the covariates are non-random, that is, the matrix X is non-random.

Suppose that the errors $e_1, \ldots, e_n$ are i.i.d. from some unknown distribution $\mathbb{F}$ with mean zero and finite variance σ^2, which is also unknown. We assume that X is of full column rank, and that

$$n^{-1} X^T X \to V \text{ as } n \to \infty \tag{3.7}$$

where V is positive definite. Under these conditions we have that

$$n^{1/2}(\hat{\beta} - \beta) \xrightarrow{\mathcal{D}} N_p\left(0, \sigma^2 V^{-1}\right). \tag{3.8}$$

See for example, Freedman (1981) where (among other places) similar results are presented and discussed. He also discussed the results for the case where the covariates are random.

We may also obtain the CLT based approximation of the distribution $\mathcal{L}_{T_n}$ of the Studentized statistic:

$$T_n = \hat{\sigma}^{-1}(X^T X)^{1/2}(\hat{\beta} - \beta), \tag{3.9}$$

for which we have

$$\mathcal{L}_{T_n} \xrightarrow{\mathcal{D}} N_p(0, \mathbb{I}_p). \tag{3.10}$$

Variations of the technical conditions are also possible. For example, Shao and Tu (1995) discuss the asymptotic normality of $\hat{\beta}$ using the conditions

$$X^T X \to \infty, \text{ and } \max_i x_i^T (X^T X)^{-1} x_i \to \infty \text{ as } n \to \infty,$$

$$\mathbb{E}|e_1|^{2+\delta} < \infty \text{ for some } \delta > 0. \tag{3.11}$$

Now consider another estimator $\hat{\beta}_{(LAD)}$ of β obtained by minimizing

$$\sum_{i=1}^{n} |Y_i - x_i\beta|.$$

In the language of Chapter 1, this may be termed an M_1-estimate in a non-i.i.d. situation. Among other places, asymptotics for this estimator may be found in Pollard (1991).

Assume that the errors are i.i.d. from some distribution $\mathbb{F}$ with a median zero, and a continuous positive density $f(\cdot)$ in a neighborhood of zero. Also assume that $n^{-1}X^T X \to V$ as $n \to \infty$ where V is positive definite. Then

$$2f(0)(X^T X)^{-1/2}(\hat{\beta}_{(LAD)} - \beta) \xrightarrow{\mathcal{D}} N(0, \mathbb{I}_p).$$

Resampling methods come to our crucial aid in such situations. Indeed we will have a problem of plenty. Since the observables are not i.i.d., we shall see that there are several eminently reasonable ways of performing resampling in this model, depending on the goal of the statistician and restrictions on the model. Resampling methods in this simple non-i.i.d. model will serve as a stepping stone to resampling in models that are more complicated and that allow for dependence in the observables. □

To summarize, we need an estimate of the distribution of a normalized or Studentized statistic, along with an estimate of the asymptotic variance of the statistic. These in general shall depend on the parent distribution, the sample size n and of course on the nature of the statistic. Even when asymptotic normality has been proved, it is not so simple outside the realm of cases like the sample mean of Example 3.1 or the l.s.e. of Example 3.3 to handle the asymptotic variance. The case of $\hat{\beta}_{(LAD)}$ illustrates this point.

3.3 Resampling methods: the jackknife and the bootstrap

Resampling techniques are computation oriented techniques for inference, that often circumvent the issues related to the asymptotics discussed above. In particular, they bypass the analytic derivations, both for asymptotic normality and for asymptotic variance. The important aspect of these techniques is the repeated use of the original sample, by drawing sample-from-the-sample, or *resampling*, often in novel ways. All these techniques broadly fall under the umbrella of *resampling techniques* or *resampling plans*.

At the least, they offer approximations to the variance of a statistic. With some resampling methods, we shall also be able to estimate "any" feature of the distribution of a statistic, which in particular includes the variance and the entire distribution. The added bonus is that some resample approximations, in a technical sense to be made precise later, achieve more accuracy, than the traditional normal approximation in a wide variety of situations. This is true even in the simplest situation of the Studentized sample mean.

However, the applicability of any resampling scheme to a given problem is not at all obvious. The fact that a resampling technique is feasible computationally does not of course imply that the resulting approximations would be correct. We now define two notions of *resampling consistency*; the first is applicable to estimation of the variance of a statistic and the second to estimation of the entire sampling distribution of a statistic, T_n say.

Definition 3.1*(Variance consistency)*: A resampling variance estimator $\hat{V}_n$ is said to be *consistent*, for $\mathbb{V}(T_n)$ or for the asymptotic variance of T_n (say $v(T_n)$), if, conditional on the data as $n \to \infty$,

$$\hat{V}_n/\mathbb{V}(T_n) \to 1 \quad \text{or} \quad \hat{V}_n/v(T_n) \to 1 \quad \text{almost surely or in probability.}$$

If the above convergence does not hold, we say that the estimator is variance inconsistent.

Definition 3.2*(Distributional consistency)*: Suppose $\mathcal{L}_n$ is the distribution of a normalized statistic or an appropriately Studentized statistic. Let $\hat{\mathcal{L}}_n$ be its resample estimate. We say that this estimate is *consistent* if, conditional on the data as $n \to \infty$,

$$\sup_x \left| \hat{\mathcal{L}}_n(x) - \mathcal{L}_n(x) \right| \longrightarrow 0, \quad \text{almost surely or in probability.}$$

If the above convergence does not hold, we say that the estimator is distributionally inconsistent.

Note that the quantity $\sup_x \left| \hat{\mathcal{L}}_n(x) - \mathcal{L}_n(x) \right|$ defines a distance metric between the distributions $\hat{\mathcal{L}}_n$ and $\mathcal{L}_n$, and we will use this metric several times in this chapter.

Clearly, conditional on the data, $\hat{V}_n$ and $\hat{\mathcal{L}}_n$ are random objects, the randomness coming from the resampling scheme used to derive the estimate. There are myriad notions of such resampling estimates. We now proceed to introduce some of the more important and basic resampling schemes.

3.3.1   Jackknife: bias and variance estimation

It stands to reason that the problem of estimating the bias and variance of a statistic should be simpler than that of estimating its entire distribution. The jackknife was introduced to do precisely this. In this section, we present the fundamentals of the jackknife procedure.

Suppose the data set is $\mathbf{Y} = (Y_1, \ldots, Y_{i-1}, Y_i, Y_{i+1}, \ldots, Y_n)$. Note that we do not assume that Y_i are necessarily i.i.d. Suppose the goal is to estimate the bias and variance of an estimator $T_n = T(\mathbf{Y})$ of some parameter θ. Following the initial ideas of Quenouille (1949), Tukey (1958) proposed the following method, which he called the *jackknife*.

Consider the data set $\mathbf{Y}_{(i)} = (Y_1, \ldots, Y_{i-1}, Y_{i+1}, \ldots, Y_n)$ obtained by deleting the i-th data point Y_i from $\mathbf{Y}$. Let

$$T_{(i)} = T(\mathbf{Y}_{(i)}), \ i = 1, \ldots, n.$$

It is implicitly assumed that the functional form of T_n is such that all the $T_{(i)}$'s are well defined. Let us define $\bar{T} = n^{-1} \sum_{i=1}^n T_{(i)}$. The jackknife estimator of the bias $\mathbb{E}T_n - \theta$ is defined as

$$(n-1)(\bar{T} - T_n).$$

Using this, we may define the bias-reduced jackknife estimator of θ as

$$T_J = T_n - \left\{ (n-1)(\bar{T} - T_n) \right\} = nT_n - (n-1)\bar{T}$$
$$= n^{-1} \sum_{i=1}^n \left(nT_n - (n-1)T_{(i)} \right).$$

Based on this, Tukey (1958) defined the *pseudo-values* as

$$\tilde{T}_i = nT_n - (n-1)T_{(i)},$$

and conjectured that the collection of pseudo-values $\{\tilde{T}_i\}$ may be roughly considered as independent copies of the random variable $n^{1/2}T_n$. While we now know that this conjecture is valid only in a limited number of problems, it suggested an immediate estimator for $\mathbb{V}(T_n)$, namely, n^{-1} times the sample variance of the pseudo-values. Thus, the delete-1 jackknife estimate of the variance or the asymptotic variance of T_n is given by

$$\widehat{\mathbb{V}}_{nJ} = n^{-1}(n-1)^{-1} \sum_{i=1}^n \left(\tilde{T}_i - n^{-1} \sum_{j=1}^n \tilde{T}_j \right)^2. \tag{3.12}$$

An alternative expression for the same quantity is

$$\widehat{\mathbb{V}}_{nJ} = (n-1)n^{-1} \sum_{i=1}^n \left(T_{(i)} - \bar{T} \right)^2. \tag{3.13}$$

Note that leaving aside the factor $(n-1)$, this may be considered to be the variance of the empirical distribution of $T_{(i)}, 1 \le i \le n$. That is the resampling randomness here.

It is easily verified that if T_n is the sample mean as in Example 3.1, then the above estimator is the same as the traditional unbiased variance estimator given in (1.4).

It was observed by Miller (1964) that the jackknife estimator $\widehat{\mathbb{V}}_{nJ}$ is consistent when T_n is a smooth function of the observations. At the same time, it performs poorly for non-smooth estimators such as the median.

Indeed, if T_n is the sample median, $\widehat{\mathbb{V}}_{nJ}/v(T_n)$ converges in distribution to $(Y/2)^2$ where Y is a chi-square random variable with two degrees of freedom (see Efron (1982), Chapter 3), and so the jackknife estimator is variance inconsistent.

Estimating the distribution of T_n is of course a more difficult problem. Suppose the data $Y_1, \ldots, Y_n$ are i.i.d. with $\mathbb{E}Y_1 = \theta$ and $\mathbb{V}Y_1 = \sigma^2$. Consider the case when $T_n = n^{-1}\sum_{i=1}^n Y_i$, the sample mean. The empirical distribution function of centered and scaled $\{T_{(i)}, i = 1, \ldots, n\}$, given by

$$\widehat{\mathbb{F}}_J(x) = n^{-1} \sum_{i=1}^n \mathcal{I}_{\{\sqrt{n(n-1)}(T_{(i)}-T_n)\leq x\}}$$

may be considered a potential estimator of the distribution of $n^{1/2}(T_n - \theta)$, which we know converges to $N(0, \sigma^2)$ using the central limit theorem. However, Wu (1990) established that $\widehat{\mathbb{F}}_J$ converges to $N(0, \sigma^2)$ if and only if the data, $Y_1, \ldots, Y_n$ are i.i.d. $N(\theta, \sigma^2)$. Hence the above delete-1 jackknife is distributionally inconsistent.

Nevertheless, the situation can be salvaged. The delete-1 jackknife can be extended in a straightforward way. We may delete d observations at a time, $1 \leq d < n$ thus yielding the delete-d jackknife. It was shown by Wu (1990) that the delete-d jackknife yields a consistent distribution estimate if d/n remains bounded away from either zero or one and the statistic T_n has some reasonable regularity properties.

It will be seen later in Section 3.5 that the different delete-d jackknives are special cases of a suite of resampling methods called the *generalized bootstrap*. The consistency results for the various jackknives can be derived from the properties of the generalized bootstrap.

3.3.2 Bootstrap: bias, variance and distribution estimation

Statistical thinking was revolutionized with the introduction of the *bootstrap* by Efron (1979), who proposed this method in an attempt to understand the jackknife better. Gradually it has spread roots and there are now myriad variations of the original idea. In this section we present an overview of the bootstrap methodology. In Section 3.4 we present details on its implementation for the sample mean and median, and then in Section 3.5 we discuss the various ways in which the bootstrap may be applied in linear regression.

Suppose the parameter of interest is $\theta = A(\mathbb{F})$, a functional of the unknown distribution $\mathbb{F}$ from which the data $Y_1, \ldots, Y_n$ is a random sample. Let $\hat{\theta}_n = A(\mathbb{F}_n)$ be its estimator where $\mathbb{F}_n$ is the e.c.d.f., and let $V(\mathbb{F}_n)$ be an estimator of the variance of $\hat{\theta}_n$. We are interested in estimating the distribution or variance of the normalized or Studentized statistics

$$Z_n = n^{1/2}\big(A(\mathbb{F}_n) - A(\mathbb{F})\big), \quad T_n = \big[V(\mathbb{F}_n)\big]^{-1/2}\big(A(\mathbb{F}_n) - A(\mathbb{F})\big).$$

Draw a simple random sample $Y_{1b}, \ldots, Y_{nb}$ with replacement (SRSWR) from the e.c.d.f. $\mathbb{F}_n$. Let $\mathbb{F}_{nB}(x)$ be the e.c.d.f. based on $\{Y_{ib}, \ i = 1, \ldots, n\}$:

$$\mathbb{F}_{nB}(x) = n^{-1} \sum_{i=1}^{n} \mathcal{I}_{\{Y_{ib} \leq x\}}.$$

Let the *bootstrap statistic* corresponding to Z_n and T_n be respectively as

$$Z_{nb} = n^{1/2}\Big(A\big(\mathbb{F}_{nB}\big) - A\big(\mathbb{F}_n\big)\Big),$$
$$T_{nb} = \big[V(\mathbb{F}_{nB})\big]^{-1/2}\Big(A\big(\mathbb{F}_{nB}\big) - A\big(\mathbb{F}_n\big)\Big).$$

The bootstrap idea is to approximate the distribution of T_n (or Z_n) by the conditional distribution of T_{nb} (or Z_{nb}). Conditional on the data $\{Y_1, \ldots, Y_n\}$, the random component in T_{nb} (or Z_{nb}) is induced only by the scheme of an SRSWR from the data. The distribution function $\mathbb{F}_n$ is completely known and hence so is the conditional distribution of T_{nb} (Z_{nb}). We use the notation $\mathbb{P}_B$, $\mathbb{E}_B$, $\mathbb{V}_B$ to respectively denote probabilities, expectations and variances of T_{nb} (Z_{nb}) or any other statistic computed using a resample, conditional on the given data $Y_1, \ldots, Y_n$. Note that by SLLN, such probabilities and moments may be approximated to arbitrary degree of accuracy by repeated

Monte Carlo simulations.

The motivation behind this novel idea is that $\mathbb{F}_n$ is a close approximation of $\mathbb{F}$, and $\mathbb{F}_{n_B}$ relates to $\mathbb{F}_n$ in the same way that $\mathbb{F}_n$ relates to $\mathbb{F}$, so T_{nb} (Z_{nb}) is a data-driven imitation of T_n (Z_n). It works when $A(\cdot)$ is a "nice" function so that this closeness idea transfers to $A(\mathbb{F}_{n_B}), A(\mathbb{F}_n)$ and $A(\mathbb{F})$.

It is important to realize that the function $\mathbb{F}_n$ need not necessarily be the empirical distribution or $\mathbb{F}_{n_B}$ need not be generated the way described above— any close approximation $\tilde{\mathbb{F}}_n$ of $\mathbb{F}$, and some "empirical-from-$\tilde{\mathbb{F}}_n$" can replace $\mathbb{F}_n$ and $\mathbb{F}_{n_B}$ in the definition of T_{nb} (Z_{nb}). We shall see that this will result in many different types of bootstrap, specially in dependent and/or non-i.i.d. situations.

In practice, a resampling scheme is implemented by repeating a few steps a (large) number of (say B) times. Note that these steps may or may not involve a simple random sampling. For each iteration $b \in \{1, 2, \ldots, B\}$, we get a value $\hat{\theta}_{nb}$ of $\hat{\theta}_n$, and imitate the centering and scaling as earlier to get T_{nb} (or Z_{nb}). The e.c.d.f. of $T_{n1}, \ldots, T_{n_B}$ is given by

$$\widehat{\mathcal{L}_{T_{nb}}}(x) = B^{-1} \sum_{b=1}^{B} \mathcal{I}_{\{T_{nb} \leq x\}}.$$

This is the *bootstrap Monte Carlo approximation of* $\mathcal{L}_{T_n}$, the distribution function of T_n. We discuss the practical implementation of the bootstrap Monte Carlo in Chapter 5.

Over the past three and a half decades a significant part of research on bootstrap has concentrated on broadening the class of problems on which bootstrap may be applied, and on establishing that by using bootstrap, better approximations are obtainable than what are obtained using more traditional asymptotic methods. Excellent review of the bootstrap, the jackknife and other resampling plans are available in many references, including the books by Efron and Tibshirani (1993); Hall (1992); Shao and Tu (1995); Davison and Hinkley (1997). In Section 3.5 we discuss the linear regression model and provide a first glimpse of the variety of resampling plans that are available.

Efron's computational ideas received strong theoretical support soon after. In Bickel and Freedman (1981) the distributional consistency of the bootstrap was demonstrated for many statistical functionals. Further, in Singh (1981), it was established that the bootstrap approximation is often better than the classical asymptotic normal approximation, in a sense we make precise in

Section 3.4.2.

Thus bootstrap was a breakthrough in two aspects:

(i) It is a computation based technique for obtaining the distribution and other properties of a large class of statistics.

(ii) It often produces an estimate that is a better approximation than the asymptotic normal approximation.

The exposition of Efron (1979) was of fundamental importance to statisticians, since it elucidated how the bootstrap is a very powerful computational tool to estimate sampling distributions in a wide array of problems.

Curiously, there have been interesting precursors of the bootstrap, and Hall (2003) discusses some of this history. This development is tied to work by statisticians in India, and is partially documented in Hubback (1946); Mahalanobis (1940, 1944, 1945, 1946b,a).

3.4 Bootstrapping the mean and the median

3.4.1 Classical bootstrap for the mean

The data is $\mathbf{Y} = (Y_1, \ldots, Y_n)^T \in \mathbb{R}^n$ where $Y_1, \ldots, Y_n$ are i.i.d. distributed as $\mathbb{F}$ with $\mathbb{E}Y_1 = \theta$, $\mathbb{V}Y_1 = \sigma^2$. The parameter of interest is the mean θ. Consider the estimator $\hat{\theta}_n = \sum Y_i / n$. The traditional or classical *Efron's bootstrap*, often called the *naive bootstrap*, can be described as follows. Draw a SRSWR $Y_{1b}, \ldots, Y_{nb}$ from the data. Let

$$\mathbf{Y}_b = (Y_{1b}, \ldots, Y_{nb})^T \in \mathbb{R}^n$$

denote the resample vector. Note that each Y_{ib} can be *any one of the* original $Y_1, \ldots, Y_n$ with a probability $1/n$. So there may be repetitions in the $\mathbf{Y}_b$ series, and chances are high that not all of the original elements of $\mathbf{Y}$ will show up in $\mathbf{Y}_b$.

Clearly, conditional on the original data $\mathbf{Y}$, the *resample* $\mathbf{Y}_b$ is random, and if we repeat the SRSWR, we may obtain a completely different $\mathbf{Y}_b$ resample vector.

Let us define

$$\hat{\theta}_{nb} = \sum_{i=1}^{n} Y_{ib}/n, \quad \text{and} \quad Z_{nb} = n^{1/2}\left(\hat{\theta}_{nb} - \hat{\theta}_n\right),$$

and let $\mathcal{L}_{Z_{nb}}$ be the distribution of Z_{nb} given $\mathbf{Y}$. This *conditional* distribution is random but depends only on the sample $\mathbf{Y}$. Hence it can be calculated exactly when we consider all the n^n possible choices of the resample vector $\mathbf{Y}_b$. The bootstrap idea is to approximate the distribution of the normalized $Z_n = n^{1/2}(\hat{\theta}_n - \theta)$ by this conditional distribution. For use later on, we also define the *asymptotically pivotal* random variable $\tilde{Z}_n = n^{1/2}(\hat{\theta}_n - \theta)/\sigma$, and denote its exact finite sample distribution by $\mathcal{L}_{\tilde{Z}_n}$.

Note that by the CLT, $\mathcal{L}_{Z_n}$ converges to $N(0, \sigma^2)$ and $\mathcal{L}_{\tilde{Z}_n}$ converges to the parameter-free distribution $N(0, 1)$ as $n \to \infty$. Further, $\mathcal{L}_{Z_{nb}}$ is also the distribution of a standardized partial sum of (conditionally) i.i.d. random variables. Hence it is not too hard to show that as $n \to \infty$, this also converges (almost surely) to the $N(0, \sigma^2)$ distribution. One easy proof of this when the third moment is finite follows from the Berry-Esseen bound given in the next section. The fundamental bootstrap result is that,

$$\sup_x \left| \mathcal{L}_{Z_{nb}}(x) - \mathcal{L}_{Z_n}(x) \right| \to 0, \quad \text{almost surely.} \tag{3.14}$$

Suppose that σ^2 is known. From the above result, either $\mathcal{L}_{Z_{nb}}$ or the CLT-based $N(0, \sigma^2)$ distribution may be used as an approximation for the unknown sampling distribution $\mathcal{L}_{Z_n}$ for obtaining confidence intervals or conducting hypothesis tests. Unfortunately, it turns out that the accuracy of the approximation (3.14) is the same as that of the normal approximation (3.1). Thus apparently no gain has been achieved.

In a vast number of practical problems and real data applications the variance σ^2 is unknown, and we now consider that case. In the classical frequentist statistical approach, we may obtain a consistent estimator of σ^2, say $\hat{\sigma}^2$, and use it as a plug-in quantity for eventual inference. An unbiased estimator of σ^2 is

$$\hat{\sigma}_u^2 = (n-1)^{-1} \sum_{i=1}^{n} \left(Y_i - \hat{\theta}_n\right)^2,$$

and in imitation of the asymptotically pivotal quantity $\tilde{Z}_n$ defined above, we now define the Studentized random variable

$$T_n = n^{1/2}\left(\hat{\theta}_n - \theta\right)/\hat{\sigma}_u.$$

This is also asymptotically pivotal, and we denote its distribution by $\mathcal{L}_{T_n}$. Readers will easily recognize this as the *t-statistic*; and when $Y_1, \ldots, Y_n$ are i.i.d. $N(\theta, \sigma^2)$, we know that $\mathcal{L}_{T_n}$ is **exactly** Student's t-distribution with $n - 1$ degrees of freedom.

However, in a vast number of practical problems and real data applications, there is little reason to believe the data to be i.i.d. $N(\theta, \sigma^2)$. Consequently, we make use of the fact that

$$\mathcal{L}_{T_n} \xrightarrow{D} N(0,1)$$

for inference. For example, a classical one-sided $1 - \alpha$ confidence interval for θ may be based on the random variable T_n and the limit of $\mathcal{L}_{T_n}$, and can thus be

$$(-\infty, \hat{\theta}_n + n^{-1/2}\hat{\sigma}_u z_{1-\alpha}), \tag{3.15}$$

where $z_{1-\alpha}$ is the $(1 - \alpha)$-th quantile of the standard Normal distribution.

Instead of the standard Normal quantile, a $t_{df=n-1}$ is often used, which makes little practical difference when n is large, yet accommodates the hope of being *exact* in case the data is i.i.d. $N(\theta, \sigma^2)$. However, questions remain about how accurate are intervals like (3.15). We will address these issues in the next few pages.

In the bootstrap approach, we may directly estimate the variance of $\hat{\theta}_{nb}$ conditional on the data. A small computation yields that this bootstrap variance estimator is $n^{-1}\hat{\sigma}_n^2$; where

$$\hat{\sigma}_n^2 = n^{-1}\sum_{i=1}^{n}\left(Y_i - \hat{\theta}_n\right)^2,$$

which is the well-known non-parametric m.l.e. of σ^2. It is easy to see as well that the bootstrap variance estimator is variance consistent according to Definition 3.1.

The corresponding bootstrap statistic is

$$T_{nb} = n^{1/2} \left(\hat{\theta}_{nb} - \hat{\theta}_n \right) / \hat{\sigma}_n,$$

and we denote its distribution by $\mathcal{L}_{T_{nb}}$. A major discovery in the early days of the bootstrap, which greatly contributed to the flourishing of this topic, is that $\mathcal{L}_{T_{nb}}$ can be a *better* approximation for $\mathcal{L}_{\tilde{Z}_n}$ compared to $N(0,1)$. This in turn leads to the fact that a bootstrap-based one-sided $(1 - \alpha)$ confidence interval for θ can be *orders of magnitude* more accurate than (3.15). We discuss these aspects in greater detail in Section 3.4.2.

We now briefly discuss the computation aspect of this approach. Note that there are n^n possible values of $\mathbf{Y}_b$. See Hall (1992), Appendix I for details on distribution of possible repetitions of $\{Y_i\}$ in $\mathbf{Y}_b$. Hence finding the exact conditional distribution of Z_{nb} involves evaluating it for all these values. This is computationally infeasible even when n is moderate, hence a Monte Carlo scheme is regularly used. Suppose we repeat the process of getting $\mathbf{Y}_b$ several times $b = 1, 2, \ldots, B$, and get $\hat{\theta}_{nb}$ and Z_{nb} for $1 \le b \le B$. Define the e.c.d.f. of these $Z_{n1}, \ldots, Z_{nB}$:

$$\widehat{\mathcal{L}}_{Z_{nb}}(\cdot) = B^{-1} \sum_{b-1}^{B} \mathcal{I}_{\{Z_{nb} \le \cdot\}}. \tag{3.16}$$

Note that this empirical distribution is easily computable, and it serves as an estimate for the true bootstrap distribution of Z_{nb}. How does it behave as $B \to \infty$? The answer is provided by the *Glivenko-Cantelli lemma*. Let $\mathbb{F}$ be any distribution function and let $\mathbb{F}_n$ be the empirical distribution based on i.i.d. observations from $\mathbb{F}$. Then this lemma says that

$$\sup_x |\mathbb{F}_n(x) - \mathbb{F}(x)| \to 0 \text{ almost surely.}$$

By an application of the Glivenko-Cantelli lemma (see Wasserman (2006), page 14), for any fixed n,

$$\lim_{B \to \infty} \sup_x |\widehat{\mathcal{L}}_{Z_{nb}}(x) - \mathcal{L}_{Z_{nb}}(x)| = 0 \text{ almost surely.}$$

Thus, $\widehat{\mathcal{L}}_{Z_{nb}}$ can be used as an approximation to $\mathcal{L}_{Z_{nb}}$, and consequently for $\mathcal{L}_{Z_n}$ as well.

Indeed we can quantify how good this approximation is. This is based on a much stronger result than the Glivenko-Cantelli lemma. The *Dvoretzky-Kiefer-Wolfowitz (DKW)* inequality says that:

$$\mathbb{P}\left[\sup_x |\mathbb{F}_n(x) - \mathbb{F}(x)| > \epsilon\right] \le 2\exp\left(-2n\epsilon^2\right) \text{ for any } \epsilon > 0. \qquad (3.17)$$

For a proof of this inequality, see Massart (1990).

Using this inequality, it can be shown that

$$\sup_x |\widehat{\mathcal{L}}_{Z_{nb}}(x) - \mathcal{L}_{Z_{nb}}(x)| = O_P(B^{-1/2}), \text{ (conditional on) } \mathbf{Y} \text{ a. s.} \qquad (3.18)$$

This quantifies how closely $\widehat{\mathcal{L}}_{Z_{nb}}$ approximates $\mathcal{L}_{Z_{nb}}$. We may choose B as large as it is computationally feasible or desirable. The choice of B in the context of confidence intervals has been studied in detail in Hall (1986).

3.4.2 Consistency and Singh property

We continue our discussion on the sample mean $\hat{\theta}_n = \sum_{i=1}^n Y_i/n$ from Example 3.1. As argued earlier, $\mathcal{L}_{Z_n}$ and $\mathcal{L}_{Z_{nb}}$ both converge to $N(0, \sigma^2)$, while $\mathcal{L}_{\tilde{Z}_n}$, $\mathcal{L}_{T_n}$ and $\mathcal{L}_{T_{nb}}$ all converge to the parameter-free $N(0, 1)$ distribution. It is not clear what is gained by using $\mathcal{L}_{Z_{nb}}$ rather than $N(0, \sigma^2)$ as an estimate of the distribution of $\mathcal{L}_{Z_n}$, or by using $\mathcal{L}_{T_{nb}}$ instead of $N(0, 1)$ or $\mathcal{L}_{T_n}$ as an estimate of $\mathcal{L}_{\tilde{Z}_n}$. To investigate this issue, we need to delve deeper.

Our starting point is the so-called Berry-Esseen bounds. Broadly speaking a Berry-Esseen bound provides an upper bound for the error committed in the normal approximation of the distribution of a statistic. We shall be concerned with such bounds only for the normalized sample mean Z_n and the *Studentized* sample mean T_n defined above in Section 3.4.1.

Early versions of the Berry-Esseen bounds are available in Lyapunov (1900, 1901); Berry (1941); Esseen (1942, 1945, 1956). The versions that we use are by Korolev and Shevtsova (2010); Shevtsova (2014) for Z_n and Bentkus and Götze (1996) for T_n respectively. Let $\Phi(\cdot)$ denote the standard normal distribution function.

Theorem 3.1. *(a) (Esseen (1956); Shevtsova (2014)) Suppose $Y_1, \dots, Y_n$ are i.i.d. from some distribution $\mathbb{F}$ with zero mean and finite variance σ^2. Let*

$\mu_3 = \mathbb{E}|Y_1|^3$. *Then for all $n \geq 1$ and some $C_0 \in (0.40973, 0.4756)$*

$$\sup_x \left| \mathbb{P}\left[\sum_{i=1}^{n} Y_i < \sigma x \sqrt{n} \right] - \Phi(x) \right| \leq C_0 \frac{\mu_3}{\sigma^3 \sqrt{n}}. \tag{3.19}$$

(b) (Bentkus and Götze (1996)) *Suppose $Y_1, \ldots, Y_n$ are i.i.d. from a distribution $\mathbb{F}$ with $\mathbb{E}Y_1 = 0$, $0 < \mathbb{V}Y_1 = \sigma^2 < \infty$. Let $\mu_3 = \mathbb{E}|Y_1|^3$. Define*

$$\bar{Y} = n^{-1} \sum_{i=1}^{n} Y_i,$$

$$\hat{\sigma}^2 = n^{-1} \sum_{i=1}^{n} \left(Y_i - \bar{Y} \right)^2, \quad and$$

$$T_n = \bar{Y}/\hat{\sigma}.$$

Then there exists an absolute constant $C > 0$ such that for all $n \geq 2$

$$\sup_x \left| \mathbb{P}\left[\sqrt{n} T_n < x \right] - \Phi(x) \right| \leq C \frac{\mu_3}{\sigma^3 \sqrt{n}}. \tag{3.20}$$

The main essence of Theorem 3.1 is that for Z_n or T_n, $n^{1/2}$ times the absolute difference between the actual sampling distribution and the Normal distribution is upper bounded by a finite constant. Thus, in using the normal approximation, we make an error of $O(n^{-1/2})$. It is also known, and can be easily verified by using i.i.d. Bernoulli variables, that the rate $n^{-1/2}$ cannot be improved in general. This puts a limit on the accuracy of the normal approximation for normalized and Studentized mean. We note in passing that such results have been obtained in many other, more complex and challenging, non-i.i.d. models and for many other statistics.

Assuming that the third moment of the distribution is finite, we can apply the Berry-Esseen bound (3.19) on Z_{nb} along with that on Z_n and this implies (3.14) mentioned earlier. However, at this point it is still not clear which is a better approximation for $\mathcal{L}_{Z_n}$; $N(0,1)$ or $\mathcal{L}_{Z_{nb}}$?

We shall now show that there is a crucial difference between dealing with normalized statistic and a Studentized statistic. Basically, for a normalized statistic there is no gain in bootstrapping. However, under suitable conditions, $\mathcal{L}_{T_{nb}}$ is a *better* estimator for $\mathcal{L}_{\bar{Z}_n}$ compared to $N(0,1)$. This is known as the *Singh property*. This property is now known to hold in many other models and statistics but we shall restrict ourselves to only T_{nb}. We need a classification of distributions to discuss this further.

Definition 3.3 *(Lattice distribution):* A random variable Y is said to have a *lattice distribution* if for some a and some $h > 0$

$$\mathbb{P}[Y = a + kh;\ k = 0, \pm 1, \pm 2, \ldots] = 1.$$

The largest such number h is called the *span* of the distribution.

Most commonly studied discrete distributions, such as the Binomial and the Poisson, are examples of lattice distributions. A distribution that does not satisfy the above property is called a non-lattice distribution. All continuous distributions are non-lattice. For the following results, recall that $\Phi(\cdot)$ and $\phi(\cdot)$ denote respectively the cumulative distribution function and the probability density function of the standard Normal distribution.

The following refinements of the Berry-Esseen theorem are needed. These are *Edgeworth expansion* results. The proofs are omitted. See Bhattacharya and Rao (1976) for a very detailed and comprehensive development of the theory of Edgeworth expansions. A selection of other relevant papers on this topic include Babu and Singh (1989a,b); Kolassa and McCullagh (1990).

Let $[y]$ denote the integer part of y. Define

$$g(y) = [y] - y + 1/2, \text{ for all } y \in \mathbb{R}.$$

Theorem 3.2. *Suppose Y_i are i.i.d. with distribution $\mathbb{F}$ that has mean 0, variance σ^2 and finite third moment μ_3.*

(a) If $\mathbb{F}$ is lattice with span h, then uniformly in x,

$$\mathcal{L}_{\tilde{Z}_n}(x) = \Phi(x) + \frac{\mu_3(1 - x^2)}{6\sigma^3 n^{1/2}}\phi(x) + \frac{h}{6\sigma^3 n^{1/2}}g(n^{1/2}\sigma h^{-1}x)\phi(x) + o(n^{-1/2}). \tag{3.21}$$

(b) If $\mathbb{F}$ is non-lattice, then uniformly in x,

$$\mathcal{L}_{\tilde{Z}_n}(x) = \Phi(x) + \frac{\mu_3(1 - x^2)}{6\sigma^3 n^{1/2}}\phi(x) + o(n^{-1/2}). \tag{3.22}$$

These bounds show that the order $O(n^{-1/2})$ in the Berry-Esseen theorem is sharp and provide additional information on the leading error term under additional conditions. Note that the nature of the leading error term is different in the lattice and non-lattice cases.

Now, if we could obtain similar expansions for $\mathcal{L}_{Z_{nb}}$ then we could use the two sets of expansions to compare the two distributions. This is a non-trivial issue. Note that for any fixed n, the bootstrap distribution of Y_{1b}, being the e.c.d.f. of $Y_1, \ldots, Y_n$, is necessarily a discrete distribution. When F is lattice (non-lattice), the bootstrap distribution may or may not be lattice (respectively non-lattice). However, it should behave as a lattice (respectively non-lattice) specially when n is large.

In an extremely remarkable work, the following result was proved by Singh (1981). For a detailed exposition on Edgeworth expansions in the context of bootstrap, see Bhattacharya and Qumsiyeh (1989), Hall (1992) and Bose and Babu (1991).

Theorem 3.3 (Singh (1981)). *Suppose Y_i are i.i.d. with distribution $\mathbb{F}$ which has mean 0, variance σ^2 and finite third moment μ_3. Then*

(a)

$$\limsup_{n \to \infty} \mathbb{E}|Y_1|^3 \sigma^{-3} n^{1/2} \sup_x |\mathcal{L}_{\tilde{Z}_n}(x) - \mathcal{L}_{T_{nb}}(x)| \leq 2C_0, \text{ almost surely,}$$

where C_0 is the constant from (a) of the Berry-Esseen Theorem.

(b) If $\mathbb{F}$ is lattice with span h, then uniformly in x, almost surely,

$$\mathcal{L}_{T_{nb}}(x) = \Phi(x) + \frac{\mu_3(1 - x^2)}{6\sigma^3 n^{1/2}} \phi(x) + \frac{h}{6\sigma^3 n^{1/2}} g(n^{1/2} \hat{\sigma}_n h^{-1} x) \phi(x) + o(n^{-1/2}).$$

Consequently

$$\limsup_{n \to \infty} n^{1/2} \sup_x |\mathcal{L}_{\tilde{Z}_n}(x) - \mathcal{L}_{T_{nb}}(x)| = \frac{h}{\sqrt{2\pi\sigma^2}}, \text{ almost surely.}$$

(c) If $\mathbb{F}$ is non-lattice, then uniformly in x, almost surely,

$$\mathcal{L}_{T_{nb}}(x) = \Phi(x) + \frac{\mu_3(1 - x^2)}{6\sigma^3 n^{1/2}} \phi(x) + o(n^{-1/2}).$$

Consequently,

$$n^{1/2} \sup_x |\mathcal{L}_{\tilde{Z}_n}(x) - \mathcal{L}_{T_{nb}}(x)| \to 0 \text{ almost surely.}$$

Part (a) shows that the difference between $\mathcal{L}_{T_{nb}}$ and $\mathcal{L}_{\tilde{Z}_n}$ has an upper bound of order $O(n^{-1/2})$, the same as the difference between $\mathcal{L}_{Z_n}$ and the normal approximation. Thus there may be no improvement in using $\mathcal{L}_{T_{nb}}$.

Part (b) shows that when the parent distribution $\mathbb{F}$ is lattice, there is no improvement in using $\mathcal{L}_{T_{nb}}$ to approximate $\mathcal{L}_{\tilde{Z}_n}$ compared to the normal approximation.

However, part (c) is most interesting. It implies that using the asymptotically pivotal Studentized statistic T_{nb} is extremely fruitful, and the bootstrap distribution $\mathcal{L}_{T_{nb}}$ is *a better estimator* of $\mathcal{L}_{\tilde{Z}_n}$ compared to the $N(0, 1)$ approximation.

This is the *higher order accuracy* or Singh Property. More complex formulations and clever manipulations can result in even higher order terms being properly emulated by the bootstrap, see Abramovitch and Singh (1985). Moreover, such higher order accuracy results have been proved in many other set ups, including Studentized versions of various U-statistics. The use of Edgeworth expansions in the context of bootstrap has been explored in details in Hall (1992), where corresponding results for many (asymptotically) pivotal random variables of interest may be found.

The sharper approximation has direct consequence in inference. Consider the problem of getting a one-sided $(1-\alpha)$ confidence interval of θ, based on the data $Y_1, \ldots, Y_n$ from some unknown distribution $\mathbb{F}$ with mean θ. We assume that $\mathbb{F}$ is non-lattice, and that the variance σ^2 is known. We discuss only the *unbounded left-tail* version, where the interval is of the form $(-\infty, R_{n,\alpha})$ for some statistic $R_{n,\alpha}$. The CLT-based estimator for this is $(-\infty, \hat{\theta}_n + n^{-1/2}\sigma z_{1-\alpha})$. From (3.22), we can compute that this interval has a $O(n^{-1/2})$ coverage error, that is

$$\mathbb{P}\left[\theta \in \left(-\infty, \hat{\theta}_n + n^{-1/2}\sigma z_{1-\alpha}\right)\right] = 1 - \alpha + O(n^{-1/2}).$$

On the other hand, using Theorem 3.3(c), if $t_{\alpha,b}$ is the α-th quantile of T_{nb}, that is, if $\mathbb{P}\left[T_{nb} \leq t_{\alpha,b}\right] = \alpha$, we have

$$1 - \alpha + O(n^{-1}) = \mathbb{P}\left[\tilde{Z}_n \geq t_{\alpha,b}\right]$$
$$= \mathbb{P}\left[\hat{\theta}_n - \theta \geq \left(n^{-1/2}\sigma t_{\alpha,b}\right]$$
$$= \mathbb{P}\left[\theta \in \left(-\infty, \hat{\theta}_n - n^{-1/2}\hat{\sigma}t_{\alpha,b}\right)\right].$$

Thus, we obtain that the bootstrap-based confidence interval $\left(-\infty, \hat{\theta}_n - n^{-1/2}\hat{\sigma}t_{\alpha,b}\right)$ is $O(n^{-1})$ accurate almost surely.

The above discussion was for the case of one-sided confidence intervals, when σ is known. Results are also available for the case when σ is unknown, and it can shown that the interval in (3.15) has a coverage error of $O(n^{-1/2})$, while the corresponding bootstrap interval has $O(n^{-1})$ coverage error. Similar results are available for two-sided intervals both when σ is known and when it is unknown. The accuracy of the coverages are different from the one sided case but the accuracy of the bootstrap intervals is still a $n^{-1/2}$ factor higher than the traditional intervals. We do not discuss the details here since several technical tools need to be developed for that. Many such details may be found in Hall (1992).

It can be seen from the above discussion that the existence and ready usability of an asymptotically pivotal random variable is critically important in obtaining the higher-order accuracy of the bootstrap estimator. Thus, Studentization is typically a crucial step in obtaining the Singh Property. Details on the Studentization and bootstrap-based inference with the Singh property is given in Babu and Singh (1983, 1984, 1985); Hall (1986, 1988) and in several other places.

3.4.3   Classical bootstrap for the median

Suppose the data at hand is $Y_1, \ldots, Y_n$ i.i.d. $\mathbb{F}$. Let ξ denote the population median and $\hat{\xi}_n$ denote the sample median. In this case, Efron's classical bootstrap is implemented as follows. Using the SRSWR scheme, generate B i.i.d. resamples $\mathbf{Y}_b = (Y_{1b}, \ldots, Y_{nb})^T \in \mathbb{R}^n$ for $b = 1, \ldots, B$. Compute the median $\hat{\xi}_{nb}$ from each of these resamples. Let

$$\widehat{\mathbb{V}}_{nJ} = B^{-1} \sum_{b=1}^{B} \left(\hat{\xi}_{nb} - \hat{\xi}_n \right)^2,$$

be the resample variance.

Under the very mild condition that $\mathbb{E}|Y_1|^\alpha < \infty$ for some $\alpha > 0$, Ghosh et al. (1984) showed that $\widehat{\mathbb{V}}_{nJ}$ is consistent for the asymptotic variance of $\hat{\xi}_n$. That is

$$4f^2(\xi)\widehat{\mathbb{V}}_{nJ} \to 1 \text{ almost surely.}$$

This bootstrap can also be used to estimate the entire sampling distribution of properly centered and scaled $\hat{\xi}_n$. Let

$$\mathcal{L}_n(x) = \mathbb{P}\Big[n^{1/2}(\hat{\xi}_n - \xi) \leq x\Big],$$

and recall that

$$\mathcal{L}_n \xrightarrow{\mathcal{D}} N(0, 1/(4f^2(\xi)))$$

as $n \to \infty$. The classical bootstrap approximation of $\mathcal{L}_n(\cdot)$ is given by

$$\hat{\mathcal{L}}_n(x) = \mathbb{P}\Big[n^{1/2}(\hat{\xi}_{nb} - \hat{\xi}_n) \leq x\Big]$$

conditional on the data. To show that $\hat{\mathcal{L}}_n$ is consistent for estimating $\mathcal{L}_n$, we use the following result:

Theorem 3.4 (Theorem 2 of Singh (1981)). *If $\mathbb{F}$ has a bounded second derivative in a neighborhood of the population median ξ and $f(\xi) > 0$, then there is a constant $C_{\mathbb{F}}$, such that*

$$\lim_{n \to \infty} \sup n^{1/4}(\log \log n)^{1/2} \sup_x |\hat{\mathcal{L}}_n(x) - \mathcal{L}_n(x)| = C_{\mathbb{F}} \quad almost\ surely.$$

A similar result is of course true for any quantile between 0 and 1.

Theorem 3.4 implies

$$\sup_x |\hat{\mathcal{L}}_n(x) - \mathcal{L}_n(x)| \longrightarrow 0 \quad almost\ surely.$$

However, note that the accuracy of the above approximation is of the order $O(n^{-1/4}(\log \log n)^{-1/2})$, which is quite low. Other resampling schemes have been studied in this context in Falk and Reiss (1989); Hall and Martin (1991); Falk (1992). We omit the details here.

3.5 Resampling in simple linear regression

Consider the linear regression model discussed earlier in Example 3.3. We now demonstrate how to conduct inference on θ using resampling plans in this model. As discussed earlier, this involves estimating the distribution function $\mathcal{L}_{T_n}$ or $\mathcal{L}_{Z_n}$, but we shall concentrate only on $\mathcal{L}_{T_n}$ for brevity, where T_n is given in (3.9).

There are several potentially applicable resampling methods in this model. Not all of them lead to consistent estimators in all situations. In this section, we present several resampling techniques that are in popular use. For simplicity and clarity, we shall restrict ourselves to the simple linear regression model (3.3) for the moment. However, many of these techniques can be adapted to more complicated models. We use the notation and definitions from Example 3.3 discussed earlier.

3.5.1   Residual bootstrap

Let us assume that the errors are i.i.d. with mean 0 and finite variance σ^2. Recall that the *residuals* from the simple linear regression are

$$r_i = Y_i - \hat{\beta}_1 - \hat{\beta}_2 x_i.$$

Bearing in mind that under mild conditions given in Example 3.3, $\hat{\beta} \to \beta$, either almost surely or in probability, it makes intuitive sense to treat the residuals as *approximately i.i.d.* Thus in this scheme, we draw the resamples from the residuals. This feature leads to the name *residual bootstrap*.

For every $b \in \{1, \ldots, B\}$, we obtain $\{r_{ib}, i = 1, \ldots, n\}$ i.i.d. from the e.c.d.f. of $r_1, \ldots, r_n$. That is, they are an SRSWR from residuals, and each r_{ib} is any one of the original residuals $r_1, \ldots, r_n$ with probability $1/n$. We then define

$$Y_{ib} = \hat{\beta}_1 + \hat{\beta}_2 x_i + r_{ib}, \quad i = 1, \ldots, n.$$

A slight modification of the above needs to be done for the case where the linear regression model is fitted without the intercept term β_1. In that case, define $\bar{r} = n^{-1} \sum_{i=1}^{n} r_i$ and for every $b \in \{1, \ldots, B\}$, we obtain $\{r_{ib}, i = 1, \ldots, n\}$ as an i.i.d. sample from $\{r_i - \bar{r}, i = 1, \ldots, n\}$. That is, they are an SRSWR from the centered residuals. Note that when an intercept term is present in the model, $\bar{r} = 0$ almost surely and hence no centering was needed.

Then for every $b = 1, \ldots, B$, we obtain the bootstrap $\hat{\beta}_b$ by minimizing

$$\sum_{i=1}^{n} \left(Y_{ib} - \beta_1 - \beta_2 x_i\right)^2.$$

Suppose $r_{ib} = Y_{ib} - \hat{\beta}_{b1} - \hat{\beta}_{b2} x_i$ are the residuals at the b-th bootstrap step.

We define the residual bootstrap noise variance estimator as

$$\hat{\sigma}_b^2 = (n-2)^{-1} \sum_{i=1}^{n} (Y_{ib} - \hat{\beta}_1 - \hat{\beta}_2 x_i)^2.$$

This is similar to the noise variance estimator $\hat{\sigma}^2$ based on the original data.

In order to state the results for the distributional consistency of the above procedure, we use a measure of distance between distribution functions. Suppose $\mathcal{F}_{r,p}$ is the space of probability distribution functions on $\mathbb{R}^p$ that have finite r-th moment, $r \geq 1$. That is,

$$\mathcal{F}_{r,p} = \left\{ G : \int_{x \in \mathbb{R}^p} ||x||^r dG(x) < \infty \right\}.$$

For two probability distributions $\mathbb{H}$ and $\mathbb{G}$ on $\mathbb{R}^p$ that belong to $\mathcal{F}_{r,p}$, the *Mallow's distance* is defined as

$$\rho_r(\mathbb{H}, \mathbb{G}) = \inf_{T_{X,Y}} \left\{ \mathbb{E}||X - Y||^r \right\}^{1/r},$$

where $T_{X,Y}$ is the collection of all possible joint distributions of (X, Y) whose marginal distributions are $\mathbb{H}$ and $\mathbb{G}$ respectively (Mallows, 1972). In a slight abuse of notation, we may also write the above as $\rho_r(X, Y)$.

Consider either the normalized residual bootstrap statistic

$$Z_{nb} = n^{1/2}(\hat{\beta}_b - \hat{\beta})$$

or the Studentized statistic

$$T_{nb} = \hat{\sigma}_b^{-1} (X^T X)^{1/2} (\hat{\beta}_b - \hat{\beta}).$$

We assume that the conditions for weak convergence of the distribution of $\hat{\beta}$ to a Gaussian distribution hold. Using the Mallow's distance as a major intermediate tool, Freedman (1981) proved that as $n \to \infty$ the distribution of Z_{nb}, conditional on the observed data, converges almost surely to $N_p(0, \sigma^2 V^{-1})$, where V is defined in (3.7). Similarly, the distribution of T_{nb}, conditional on the observed data, converges almost surely to the standard Normal distribution on $\mathbb{R}^p$ denoted by $N_p(0, \mathbb{I}_p)$. That is, as $n \to \infty$ almost surely

$$\mathcal{L}_{Z_{nb}} \to N_p(0, \sigma^2 V^{-1}), \quad \text{and} \quad \mathcal{L}_{T_{nb}} \to N_p(0, \mathbb{I}_p).$$

Coupled with the fact that the normalized and Studentized statistic that were formed using the original estimator $\hat{\beta}$ converge to the same limiting distributions (see (3.8), (3.10)), this shows that the residual bootstrap is consistent for both T_n and Z_n. In practice, a Monte Carlo approach is taken, and the e.c.d.f of the $\{Z_{nb}, b = 1, \ldots, B\}$ or $\{T_{nb}, b = 1, \ldots, B\}$ are used for bootstrap inference with a choice of (large) B.

3.5.2   Paired bootstrap

Here, we consider the data $\{(Y_i, x_i), i = 1, \ldots, n\}$ as consisting of n *pairs of observations* (Y_i, x_i), and we draw an SRSWR from these data-pairs. Thus, for every $b \in \{1, \ldots, B\}$, we obtain $\{(Y_{ib}, x_{ib}), i = 1, \ldots, n\}$, where each (Y_{ib}, x_{ib}) is any one of the original sampled data $(Y_1, x_1), \ldots, (Y_n, x_n)$ with probability $1/n$. This is a natural resampling technique to use when the regressors $\{x_i\}$ are random variables.

Once the resample $\{(Y_{ib}, x_{ib}), i = 1, \ldots, n\}$ is obtained, we use the simple linear regression model, and obtain $\hat{\beta}_{nb}$ by minimizing

$$\sum_{i=1}^{n} (Y_{ib} - \beta_1 - \beta_2 x_{ib})^2 .$$

This bootstrap is known as the *paired bootstrap*. Assume that (Y_i, x_i) are i.i.d. with $\mathbb{E}\|(Y_i, x_i)\|^4 < \infty$, $\mathbb{V}(x_i) \in (0, \infty)$, $\mathbb{E}(e_i|x_i) = 0$ almost surely. Let the distribution of $T_{nb} = \hat{\sigma}_b^{-1}(X^T X)^{1/2}(\hat{\beta}_b - \hat{\beta})$, conditional on the data, be $\mathcal{L}_{T_{nb}}$. Freedman (1981) proved that $\mathcal{L}_{T_{nb}} \to N_p(0, \mathbb{I}_p)$ almost surely. This establishes the distributional consistency of the paired bootstrap for T_n.

It can actually be shown that under very standard regularity conditions, the paired bootstrap is distributionally consistent even in many cases where the explanatory variables are random, and when the errors are heteroscedastic. It remains consistent in multiple linear regression even when p varies with the sample size n and increases as the sample size increases. These details follow from the fact that the paired bootstrap is a special case of the *generalized bootstrap* described later, for which corresponding results were established in Chatterjee and Bose (2005). However, there are additional steps needed before the Singh property can be claimed for this resampling scheme.

Note that the paired bootstrap can be computationally expensive, since it involves repeated inversions of matrices. Also, the SRSWR sampling scheme may produce resamples where the design matrix may not be of full column rank. However, such cases happen with exponentially small probability (Chatterjee and Bose, 2000), and may be ignored during computation.

3.5.3 Wild or external bootstrap

In this scheme, we obtain the resamples using the *residuals* $\{r_i\}$ together with an i.i.d. draw $Z_{1b}, \ldots, Z_{nb}$ from some known distribution $\mathbb{G}$ which has mean 0 and variance 1. There could be more restrictions on $\mathbb{G}$, for example, we might want its skewness to be a constant that we specify. We then define

$$r_{ib} = Z_{ib} r_i, i = 1, \ldots, n, \text{ and}$$
$$Y_{ib} = \hat{\beta}_1 + \hat{\beta}_2 x_i + r_{ib}, i = 1, \ldots, n.$$

We obtain $\hat{\beta}_b$ by minimizing

$$\sum_{i=1}^{n} (Y_{ib} - \beta_1 - \beta_2 x_i)^2.$$

This is known as the *wild* or the *external* bootstrap. Under (3.11), Shao and Tu (1995) established the distributional consistency of this resampling scheme.

When random regressors are in use, Mammen (1993) established the distributional consistency for both the paired and the wild bootstrap under very general conditions.

3.5.4 Parametric bootstrap

This resampling scheme is an alternative when $\mathbb{F}$ is known, except possibly for some finite-dimensional parameter ξ, which can be estimated from the available data by $\hat{\xi}_n$. For example, in the Gauss-Markov model, $\mathbb{F} = N(0, \sigma^2)$, and thus $\xi \equiv \sigma^2$.

For every $b \in \{1, \ldots, B\}$, we obtain $Z_{1b}, \ldots, Z_{nb}$ i.i.d. with distribution $\mathbb{F}(\cdot; \hat{\xi}_n)$, where $\mathbb{F}(\cdot; \hat{\xi}_n)$ is the distribution $\mathbb{F}$ with $\hat{\xi}_n$ in place of the unknown parameter ξ. For example, in the Gauss-Markov model, $\mathbb{F}(\cdot; \hat{\xi}_n)$ is $N(0, \widehat{\sigma^2})$.

We then define

$$Y_{ib} = \hat{\beta}_1 + \hat{\beta}_2 x_i + Z_{ib}, i = 1, \ldots, n.$$

We obtain $\hat{\beta}_b$ by minimizing

$$\sum_{i=1}^{n} \left(Y_{ib} - \beta_1 - \beta_2 x_i\right)^2.$$

The parametric bootstrap is distributionally consistent under the regularity properties listed in (3.11).

3.5.5   Generalized bootstrap

While the above schemes differed in how $Y_{ib}, i = 1, \ldots, n$ are generated, notice that the basic feature of obtaining estimators and their resample-based versions by use of minimization of the sum of squared error, that is, *ordinary least squares* (OLS) methodology is retained. The material below constitute a special case of the results of Chatterjee and Bose (2005).

In *generalized bootstrap* (GBS in short) we dispense with generation of $Y_{ib}, i = 1, \ldots, n$ altogether. For every $b \in \{1, \ldots, B\}$, we obtain $\hat{\beta}_b$ directly by minimizing the *weighted least squares* (WLS)

$$\sum_{i=1}^{n} \mathbb{W}_{ib} \left(Y_i - \beta_1 - \beta_2 x_i\right)^2.$$

Here $\{\mathbb{W}_{1b}, \ldots, \mathbb{W}_{nb}\}$ are a set of *random weights*, and the properties of the GBS are entirely controlled by the distribution of the n-dimensional vector $\mathcal{W}_{nb} = (\mathbb{W}_{1b}, \ldots, \mathbb{W}_{nb})$. We discuss special cases below, several of which were first formally listed in Præstgaard and Wellner (1993). We omit much of the details, and just list the essential elements of the resampling methodology. Let Π_n be the n-dimensional vector all of whose elements are $1/n$, thus $\Pi_n = \frac{1}{n}\mathbf{1}_n \in \mathbb{R}^n$ where $\mathbf{1}_n$ is the n-dimensional vector of all 1's.

1. **[Paired bootstrap]** We get the paired bootstrap as a special case of the GBS by specifying that each $\mathcal{W}_{nb} = (\mathbb{W}_{1b}, \ldots, \mathbb{W}_{nb})$ is an i.i.d. draw from the Multinomial(n, Π_n) distribution.

2. **[Bayesian bootstrap]** Suppose α_n is the n-dimensional vector with all elements equal to α, for some $\alpha > 0$. We get the Bayesian bootstrap as a special case of the GBS by specifying that each $\mathcal{W}_{nb} = (\mathbb{W}_{1b}, \ldots, \mathbb{W}_{nb})$ is an i.i.d. draw from the Dirichlet(α_n) distribution. The properties of the Bayesian bootstrap is very similar to that of the paired bootstrap,

however, this one has a Bayesian interpretation.

3. [**Moon bootstrap**] (*m* out of *n* bootstrap)   We get the *m*-out-of-*n* bootstrap (which we call the moon-bootstrap) as a special case of the GBS by specifying that each $\mathcal{W}_{nb} = (\mathbb{W}_{1b}, \dots, \mathbb{W}_{nb})$ is an i.i.d. draw from the Multinomial(m, Π_n) distribution. Under the conditions $m \to \infty$ as $n \to \infty$ but $m/n \to 0$, the moon-bootstrap is consistent in a very wide class of problems where most other methods fail. This is one of the most robust resampling methods available.

For comparison with subsampling later on, let us also describe the mechanism by which the moon-bootstrap is generated classically. We fix a "subsample" size *m* where *m* is (considerably) smaller than *n*. Then for every $b \in \{1, \dots, B\}$, we obtain subsamples of size *m* $\{(Y_{ib}, x_{ib}),\ i = 1, \dots, m\}$ by drawing an SRSWR from $(Y_1, x_1), \dots, (Y_n, x_n)$.

4. [**The delete-*d* jackknife**] In the context of some variance consistency results, it was established in Chatterjee (1998), that all the jackknives are special cases of GBS. We outline the main idea behind this below.

Let $\mathbb{N}_n = \{1, 2, \dots, n\}$ and $\mathcal{S}_{n,d} = \{\text{all subsets of size } d \text{ from } \mathbb{N}_n\}$. We identify a typical element of $\mathcal{S}_{n,d}$ by $S = \{i_1, i_2, \dots, i_d\}$. There are $\binom{n}{d}$ elements in $\mathcal{S}_{n,d}$. Let $S^C = \mathbb{N}_n \setminus S$. Let $\{\xi_S, S \in \mathcal{S}_{n,d}\}$ be a collection of vectors in $\mathbb{R}^n$ defined by $\xi_S = (\xi_S(1), \xi_S(2), \dots, \xi_S(n))$, where

$$
\xi_S(i) = \begin{cases} (n-d)^{-1}n & \text{if } i \in S^C \\ 0 & \text{if } i \in S. \end{cases}
$$

Then the delete-*d* jackknife can be identified as a special case of the GBS with the resampling weight vector $\mathcal{W}_{nb} = (\mathbb{W}_{1b}, \mathbb{W}_{2b}, \dots, \mathbb{W}_{nb})$ which has the probability law

$$
\mathbb{P}[\mathcal{W}_{nb} = \xi_S] = \binom{n}{d}^{-1}, \quad S \in \mathcal{S}_{n,d}. \tag{3.23}
$$

5. [**The deterministic weight bootstrap**] This is a generalization of the delete-*d* jackknife. Suppose $\{\xi_n\}$ is a sequence of numbers. Consider a random permutation of its first *n* elements as $\mathcal{W}_{nb}$. See Præstgaard and Wellner (1993) for further conditions on the sequence.

6. [**The double bootstrap**] Suppose $\mathcal{M}_n$ follows a Multinomial (m, Π_n)

distribution, and conditional on $\mathcal{M}_n$, suppose that each of the resampling weights $\mathcal{W}_{nb} = (\mathbb{W}_{1b}, \ldots, \mathbb{W}_{nb})$ is an i.i.d. draw from the Multinomial $(n, n^{-1}\mathcal{M}_n)$ distribution. This particular resampling scheme is often used for calibration of other resampling techniques, to ensure that the Singh property holds.

7. [**The Polya-Eggenberger bootstrap**] This is a variant of the double bootstrap. Suppose $\mathcal{M}_n$ follows a Dirichlet($\boldsymbol{\alpha}_n$) distribution, where $\boldsymbol{\alpha}_n = \alpha\mathbf{1}_n \in \mathbb{R}^n$ for some $\alpha > 0$. Conditional on $\mathcal{M}_n$, the resampling weights $\mathcal{W}_n$ follow a Multinomial $((n, n^{-1}\mathcal{M}_n))$.

8. [**The hypergeometric bootstrap**] In this procedure, we have

$$\mathbb{P}\big[\mathbb{W}_{1b} = b_1, \ldots, \mathbb{W}_{nb} = b_n\big] = \binom{nK}{n}^{-1} \prod_{i=1}^{n} \binom{K}{b_i},$$

for integers $b_1, \ldots, b_n$ satisfying $0 \le b_i \le K$, $\sum_{i=1}^{n} b_i = n$.

9. [**Subsampling**] Subsampling is related to jackknives and the moon-bootstrap. This method, like the moon-bootstrap, is one of the most robust inferential methods available. Essentially, if $\mathcal{L}_{T_n}$ has a limit, subsampling produces a consistent estimate of it. In subsampling, we fix a "subsample" size m where m is (considerably) smaller than n. Then for every $b \in \{1, \ldots, B\}$, we obtain subsamples of size m $\{(Y_{ib}, x_{ib}), i = 1, \ldots, m\}$ by drawing a *simple random sample* **without** *replacement* (SRSWOR) from the original data $(Y_1, x_1), \ldots, (Y_n, x_n)$. It is crucial that the sampling should be *without* replacement, otherwise we just get the moon-bootstrap. Subsampling can be identified also with the delete-$n - m$ jackknife.

3.6   Exercises

1. Show that if T_n is the sample mean as in Example 3.1, then its jackknife variance estimate is same as the traditional unbiased variance estimator given in (1.4).

2. Show that the two expressions for the jackknife variance estimator $\widehat{\mathbb{V}}_{nJ}$ given in (3.12) and (3.13) are identical.

3. Show that if T_n is the sample mean as in Example 3.1, then its naive bootstrap variance estimate is

$$n^{-2} \sum_{i=1}^{n} (Y_i - \hat{\theta}_n)^2.$$

4. Verify that there are n^n possible values of $\mathbb{Y}_b$ in Efron's bootstrap.

5. Consider Efron's classical bootstrap using the SRSWR drawing from the data.

   (a) What is the probability that at least one of the original values $Y_1, Y_2, \ldots, Y_n$ does not show up in the resample values $Y_{1b}, \ldots, Y_{nb}$?

   (b) Compute for $k = 1, \ldots, n$, the probability that exactly k of the original values $Y_1, Y_2, \ldots, Y_n$ show up in the collection of resample values $Y_{1b}, \ldots, Y_{nb}$.

   (c) Comment on the cases $k = 1$ and $k = n$.

6. Look up the proof of D-K-W inequality (3.17).

7. Use the D-K-W inequality to prove the Gilvenko-Cantelli Lemma.

8. Show how claim (3.18) follows from the D-K-W inequality.

Chapter 4

Resampling U-statistics and M-estimators

4.1   Introduction

Recall from Chapter 1 that if $Y_1, \ldots, Y_n$ is an i.i.d. sample from some probability distribution function $\mathbb{F}$, and $\theta = \mathbb{E}h(Y_1, Y_2, \ldots, Y_m)$ where $h(\cdot)$ is symmetric in its arguments, then the U-statistic,

$$U_n = \binom{n}{m}^{-1} \sum_{1 \le i_1 < i_2 < \ldots < i_m \le n} h(Y_{i_1}, \ldots, Y_{i_m})$$

is the usual estimator for θ. In order to perform hypothesis tests or to obtain confidence intervals for θ, we need the exact or approximate distribution of U_n. There are no closed form exact distribution of U-statistics in general. We established the asymptotic normality of U-statistics in Theorem 1.1 of Chapter 1. However, the asymptotic variance is typically unknown and has to be estimated. Similarly, the M_m-estimators discussed in Chapter 2 have an asymptotic normal distribution under fairly general conditions, but the asymptotic variance is typically parameter-dependent and has to be estimated.

An alternative to using such asymptotic distributional results and estimated parameters is to use resampling, which has been introduced in Chapter 3. We have seen in Chapter 3 that in addition to being usable for

© Springer Nature Singapore Pte Ltd. 2018 and Hindustan Book Agency 2018
A. Bose and S. Chatterjee, *U-Statistics, Mm-Estimators and Resampling*, Texts
and Readings in Mathematics 75, https://doi.org/10.1007/978-981-13-2248-8_4

estimating asymptotic variance, many resampling methods can provide direct approximations to the sampling distribution of statistics of interest, and thus circumvent the need for using asymptotic variance altogether. Such resampling-based distribution estimators may have the very attractive Singh property (higher order accuracy) described in Chapter 3, and thus they can be better than CLT-based asymptotic approximations.

In this chapter, we discuss some resampling techniques that are applicable for estimating sampling distributions and asymptotic variances of U-statistics and M_m-estimators. We will discuss how the GBS estimators discussed in Chapter 3 arise naturally in this context.

First in Section 4.2 we review some of the more traditional resampling schemes as applied to U-statistics. We will see that a naive application of Efron's bootstrap or a jackknife technique may lead to poorly defined estimates. Consequently there is a need for more careful application of resampling schemes in this context. In Section 4.3 we present the GBS for U-statistics, and show how some of the classical resampling methods can be treated as special cases of the GBS. Moreover, the very general form of the GBS suggests a variety of resampling weights that may be used in this context. We discuss an additive weights resampling technique in Section 4.4, which is also a special case of the GBS, but deserves special attention owing to the potential of savings in the computational requirements. Then in Section 4.5 we present the GBS for M_m-estimators, of which the U-statistics are a special case.

Let U_{nb} be the generic form of any resampled version of U_n. Define

$$\mathcal{L}_{U_n}(x) = \mathbb{P}\big[n^{1/2}(U_n - \theta) \le x\big],$$
$$\mathcal{L}_{U_{nb}}(x) = \mathbb{P}\big[n^{1/2}m\tau_n^{-1}(U_{nb} - U_n) \le x \big| Y_1, \ldots, Y_n\big],$$

where τ_n is some appropriate scaling.

Recall that the corresponding resampling scheme is said to be *distributionally consistent* if

$$\sup_{x \in \mathbb{R}}\big|\mathcal{L}_{U_n}(x) - \mathcal{L}_{U_{nb}}(x)\big| \to 0$$

either in probability or almost surely as $n \to \infty$.

4.2   Classical bootstrap for U-statistics

Recall that in Efron's bootstrap or *multinomial bootstrap* the resample is chosen using an SRSWR of size n from the e.c.d.f. When applied to the U-statistic U_n, this bootstrap is implemented with the following steps: from the data $Y_1, \ldots, Y_n$, draw a simple random sample with replacement, say $Y_{1b}, \ldots, Y_{nb}$. Then *Efron's bootstrap U-statistics* is defined by

$$U_{nb} = \binom{n}{m}^{-1} \sum_{1 \leq i_1 < i_2 < \cdots < i_m \leq n} h(Y_{i_1 b}, \ldots, Y_{i_m b}). \qquad (4.1)$$

Conditional on the data $Y_1, \ldots, Y_n$, the distribution (or variance) of the statistic U_{nb} may be used to approximate the distribution (or variance) of U_n. Consistency results for Efron's bootstrap were obtained by Bickel and Freedman (1981) and Athreya et al. (1984) for normalized and Studentized U-statistics respectively. As an example, we present the following result, which is based on Bickel and Freedman (1981).

Theorem 4.1 (Bickel and Freedman (1981)). *Let $Y_1, \ldots, Y_n$ be i.i.d. random variables and let*

$$U_n = \binom{n}{m}^{-1} \sum_{1 \leq i_1 < \cdots < i_m \leq n} h(Y_{i_1}, \ldots, Y_{i_m})$$

with $m = 2$. Suppose that $\mathbb{E}|h(Y_1, Y_2)|^2 < \infty$, $\mathbb{E}|h(Y_1, Y_1)|^2 < \infty$, and $\int h(x, y) dF(y)$ is not a constant. Then the distribution of U_{nb} conditional on the data is a consistent estimator of the distribution of U_n with $\tau_n = 1$, that is, almost surely,

$$\sup_{x \in \mathbb{R}} \left| \mathcal{L}_{U_n}(x) - \mathcal{L}_{U_{nb}}(x) \right| \to 0.$$

Later Helmers (1991) proved the Singh property of the bootstrap approximation for the distribution of a Studentized non-degenerate U-statistics of degree 2. Suppose the kernel satisfies $\mathbb{E}h(Y_1, Y_2) = \theta$ and the corresponding U-statistic

$$U_n = \binom{n}{2}^{-1} \sum_{1 \leq i < j \leq n} h(Y_i, Y_j)$$

has the Hoeffding projection

$$h_1(Y_i) = \mathbb{E}\big(h(Y_i, Y_j) - \theta | Y_i\big)$$

with positive variance δ_1. Helmers (1991) defined

$$S_n^2 = 4(n-1)(n-2)^{-2} \sum_{i=1}^{n} \Big((n-1)^{-1} \sum_{j=1}^{n} h(Y_i, Y_j) - U_n\Big)^2,$$

and proposed $n^{-1}S_n^2$ as a jackknife-type estimator of the variance of U_n. Define

$$\mathcal{L}_{T_n}(x) = \mathbb{P}\Big[n^{1/2}S_n^{-1}(U_n - \theta) \leq x\Big], \text{ for } x \in \mathbb{R}.$$

For the bootstrap versions, we compute U_{nb} and S_{nb}^2 using the formula for U_n and S_n^2 given above but replacing Y_i's with Y_{ib}'s. Define

$$\theta_n = n^{-2} \sum_{i,j=1}^{n} h(Y_i, Y_j),$$

which is very close to U_n, except that the $h(Y_i, Y_i)$ terms are now included in the summation, which thus has n^2 terms. Consequently, θ_n has the scaling factor n^2, comparable to the $\binom{n}{2}$ factor that appears in the denominator of U_n. Using these, we obtain the bootstrap distributional estimator

$$\mathcal{L}_{T_{nb}}(x) = \mathbb{P}_B\Big[n^{1/2}S_{nb}^{-1}(U_{nb} - \theta_n) \leq x\Big], \ x \in \mathbb{R}.$$

Helmers (1991) first established a one-term Edgeworth expansion of $\mathcal{L}_{T_n}$. Then, under the additional condition that $\mathbb{E}|h(Y_1, Y_1)|^3 < \infty$, the Singh property was established. We state this result below.

Theorem 4.2 (Helmers (1991)). *Suppose that the distribution of the Hoeffding projection h_1 is non-lattice, $\mathbb{E}|h(Y_1, Y_2)|^{4+\epsilon} < \infty$ for some $\epsilon > 0$, and $\mathbb{E}|h(Y_1, Y_1)|^3 < \infty$. Then*

$$\sup_x |\mathcal{L}_{T_n}(x) - \mathcal{L}_{T_{nb}}(x)| = o(n^{-1/2}) \text{ as } n \to \infty \text{ almost surely.}$$

A general consistency result, regardless of the degree of degeneracy of a U-statistic, was obtained by Arcones and Gine (1992), as long as the degree of degeneracy is known. In a different direction, Helmers and Hušková (1994) considered multivariate U-quantiles and showed consistency in distribution estimation. A number of other authors have also studied Efron's bootstrap of U-statistics, see references in the papers mentioned above and also the book by Shao and Tu (1995). As with most other problems where resampling is used, a Monte Carlo implementation may be used for numeric computations.

The performance of Efron's bootstrap is affected by the behaviour of the kernel. Note that in this resampling, values such as $h(Y_1, Y_1, \ldots, Y_1)$ may appear in the resample while they are absent in the original definition of U_n. This may create inconsistency of the bootstrap. An example is provided in Bickel and Freedman (1981) and reproduced in Shao and Tu (1995), and we sketch it below.

Example 4.1: Supposed $m = 2$, and define $h(x, x) = \exp\{1/x\}$, and consider the population distribution to be continuous Uniform $(0, 1)$. Note that under this distribution we have $\mathbb{E}h(Y, Y) = \infty$. We write the kernel $h(x, y) = h_1(x, y) + h_2(x, y)$ where $h_1(x, y) = h(x, y)\mathcal{I}_{x \neq y}$. Let U_{n1} and U_{n2} be the U-statistics corresponding to h_1 and h_2 respectively. Define $U_n = U_{n1} + U_{n2}$, and note that since the population distribution is continuous, $U_{n2} = 0$ almost surely. Also, note that suitably centered and scaled U_n has an asymptotic normal distribution. It can be shown that if we choose to use Efron's bootstrap to estimate the distribution of U_n, we get an inconsistent result, since $n^{1/2}(U_{n2b} - U_{n2}) \xrightarrow{\mathbb{P}} \infty$. □

4.3   Generalized bootstrap for U-statistics

The kind of deficiency we saw for Efron's bootstrap can be remedied by using generalized bootstrap (GBS) instead.

We first introduce the notation for a triangular sequence of random variables, to be called the bootstrap weights in the sequel. Suppose for every positive integer n, we consider all possible m-dimensional vectors $(i_1, \ldots, i_m)$ where each $i_j \in \{1, \ldots, n\}$, and i_j's distinct. There are clearly $\binom{n}{m}$ such vectors. Corresponding to these m-dimensional vector of distinct integers, we define the triangular sequence of random variables $\mathbb{W}_{n:i_1, \ldots, i_m}$ as *bootstrap weights*. We will shortly present examples of such *bootstrap weights*, and later present properties of such weight vectors when we discuss theoretical results.

For a U-statistic with kernel $h(y_1, \ldots, y_m)$, define its *generalized bootstrap*

(GBS) version as

$$U_{nb} = \binom{n}{m}^{-1} \sum_{1 \leq i_1 < \cdots < i_m \leq n} \mathbb{W}_{n:i_1,\ldots,i_m} h(Y_{i_1},\ldots,Y_{i_m}). \qquad (4.2)$$

Example 4.2 *(Efron's bootstrap as a special case of GBS)*: Suppose $\mathbb{W}_{n:i}$ is the (random) number of times the data point Y_i is obtained when we implement Efron's bootstrap of drawing an SRSWR from the original data. It can be seen that in this case, the vector $\mathcal{W}_n = (\mathbb{W}_{n:1},\ldots,\mathbb{W}_{n:n})$ follows a *Multinomial*$(n; 1/n,\ldots,1/n)$ distribution. Define

$$\mathbb{W}_{n:i_1,\ldots,i_m} = \prod_{j=1}^{m} \mathbb{W}_{n:i_j}. \qquad (4.3)$$

Consider the case of a U-statistic with kernel of order $m = 1$. Here, it is easy to see that Efron's bootstrap is simply a GBS with the choice (4.3).

For general m, Efron's bootstrap for U-statistic (4.1) *cannot* be written like the GBS (4.2). However, relation (4.3) suggests a new way of defining the bootstrap version of a U-statistic, namely

$$U_{nb} = \binom{n}{m}^{-1} \sum_{1 \leq i_1 < i_2 < \ldots < i_m \leq n} \prod_{j=1}^{m} \mathbb{W}_{n:i_j} h(Y_{i_1},\ldots,Y_{i_m}). \qquad (4.4)$$

Hušková and Janssen (1993b) established the distributional consistency of (4.4) for U-statistics of degree $m = 2$. Later Hušková and Janssen (1993a) extended this result to degenerate U-statistics. $\qquad \square$

It is important to realize that the resampling weights used in (4.2) are of a very general form, and the choice of using a multiplicative form as in (4.4) is a further special case. Instead, for example, consider the *additive form* of resampling weights where

$$\mathbb{W}_{n:i_1,\ldots,i_m} = m^{-1} \sum_{j=1}^{m} \mathbb{W}_{n:i_j}. \qquad (4.5)$$

In Section 4.4 below we discuss the properties of this particular choice of resampling weights in detail. We will see that using the additive form (4.5) can lead to significant improvement in computational efficiency, without compromising on the accuracy.

Recall Example 2.2 from Chapter 2, where we showed that all U-statistics are M_m-estimators. That is, for any kernel $h(y_1, \ldots, y_m)$, which is symmetric in its arguments, the corresponding U-statistic can be obtained as the unique M_m-estimator when we use the contrast function

$$f(y_1 \ldots, y_m, \theta) = \big(\theta - h(y_1, \ldots, y_m)\big)^2 - \big(h(y_1, \ldots, y_m)\big)^2.$$

In view of this, instead of just presenting the analysis of GBS for U-statistics given in (4.2), we present in Section 4.5 the full discussion of GBS for generic M_m-estimators.

4.4 GBS with additive weights

Let $\{\mathbb{W}_{n:i}, \; 1 \le i \le n, \; n \ge 1\}$ be a triangular sequence of *non-negative, row-wise exchangeable* random variables, independent of $\{Y_1, \ldots, Y_n\}$. Recall the notation $\mathbb{P}_B$, $\mathbb{E}_B$, $\mathbb{V}_B$ to respectively denote probabilities, expectations and variances of these resampling weights, conditional on the given data $Y_1, \ldots, Y_n$. We henceforth drop the first suffix in the weights $\mathbb{W}_{n:i}$ and denote it by $\mathbb{W}_i$. We fix that $\mathbb{E}_B \mathbb{W}_i = 1$ for all $i = 1, \ldots, n$, and write

$$\tau_n^2 = \mathbb{V}_B \mathbb{W}_i \quad \text{and} \quad W_i = \tau_n^{-1}(\mathbb{W}_i - 1). \tag{4.6}$$

Since row-wise exchangeability of $\mathbb{W}_{n:i}$'s is assumed, we may unambiguously adopt the notation

$$c_{abc\ldots} = \mathbb{E}_B W_1^a W_2^b W_3^c \ldots.$$

Generic constants at various places will be denoted by k and K.

We will assume the following conditions on the weights:

$$\mathbb{E}_B \mathbb{W}_1 = 1, \tag{4.7}$$

$$0 < k < \tau_n^2 < K, \tag{4.8}$$

$$c_{11} = O(n^{-1}), \tag{4.9}$$

$$c_{22} \to 1, \tag{4.10}$$

$$\sup_n c_4 < \infty. \tag{4.11}$$

Note that when $\mathcal{W}_n = (\mathbb{W}_1, \ldots, \mathbb{W}_n)$ has the *Multinomial*$(n; 1/n, \ldots, 1/n)$ distribution that *links* the GBS to Efron's bootstrap, all the above conditions are satisfied. In fact for these weights,

$$\tau_n^2 = 1 - 1/n \quad \text{and} \quad c_{11} = -1/(n-1). \tag{4.12}$$

Define as usual, the centered first projection of h as

$$h_1(Y_1) = \mathbb{E}\big[h(Y_1, \ldots, Y_m)|Y_1\big] - \theta,$$

and assume that

$$\mathbb{E}h^4(Y_1, \ldots, Y_m) < \infty, \tag{4.13}$$

$$\mathbb{E}|h_1(Y_1)|^{4+\delta} < \infty, \quad \text{for some } \delta > 0, \tag{4.14}$$

$$\mathbb{E}h_1^2(Y_1) > 0. \tag{4.15}$$

Theorem 4.3. *Suppose the kernel h satisfies* (4.13)-(4.15) *and the resampling weights are of the form* (4.5) *where $\mathbb{W}_{n:i}$ satisfy* (4.7)-(4.11) *and also*

$$\sum_{i=1}^{n} \mathbb{W}_{n:i} = n.$$

Then

$$\sup_{x \in \mathbb{R}} \big|\mathcal{L}_{U_n}(x) - \mathcal{L}_{U_{nb}}(x)\big| \to 0 \quad \text{almost surely as } n \to \infty. \tag{4.16}$$

The condition $\sum_{i=1}^{n} \mathbb{W}_{n:i} = n$ and (4.8) together imply (4.9) and (4.12). A convergence in probability version of this theorem may also be established with more relaxed condition than (4.13)-(4.14).

To prove the theorem we need a CLT for weighted sums of row-wise exchangeable variables from Præstgaard and Wellner (1993). The idea behind this result is Hajek's classic CLT (Hájek (1961)) for sampling without replacement.

Theorem 4.4 (Præstgaard and Wellner (1993)). *Let $\{a_{mj}\}$ be a triangular array of constants, and let B_{mj}, $j = 1, \ldots, m$, $m \geq 1$ be a triangular array of row-exchangeable random variables such that*

$$m^{-1} \sum_{j=1}^{m} \big(a_{mj} - \bar{a}_m\big)^2 \to \sigma^2 > 0,$$

$$m^{-1} \max_{j=1,\dots,m} \left(a_{mj} - \bar{a}_m\right)^2 \to 0,$$

$$m^{-1} \sum_{j=1}^{m} \left(B_{mj} - \bar{B}_m\right)^2 \xrightarrow{\text{P}} c^2 > 0,$$

$$\lim_{K \to \infty} \limsup_{m \to \infty} \mathbb{E}\left(B_{mj} - \bar{B}_m\right)^2 \mathcal{I}_{\{|B_{mj} - \bar{B}_m| > K\}} = 0.$$

Here,

$$\bar{a}_m = m^{-1} \sum_{j=1}^{m} a_{mj} \quad and \quad \bar{B}_m = m^{-1} \sum_{j=1}^{m} B_{mj}.$$

Then

$$\frac{1}{\sqrt{m}} \sum_{j=1}^{m} \left(a_{mj} B_{mj} - \bar{a}_m \bar{B}_m\right) \xrightarrow{\mathcal{D}} N(0, c^2 \sigma^2). \tag{4.17}$$

Proof of Theorem 4.3: Let $s = (i_1, \dots, i_m)$ denote a typical multi-index. Without any scope for confusion, $\mathbb{W}_s = \mathbb{W}_{n:i_1,\dots,i_m} = m^{-1} \sum_{j=1}^{m} \mathbb{W}_{i_j}$. Let

$$W_s = \tau_n^{-1}\left(\mathbb{W}_s - 1\right),$$

$$h_s = h(Y_{i_1}, \dots, Y_{i_m}),$$

$$g_s = h_s - \sum_{j=1}^{m} h_1(Y_{i_j}) - \theta,$$

$$N = \binom{n}{m}.$$

Then

$$n^{1/2} \tau_n^{-1}\left(U_{nb} - U_n\right) = n^{1/2} N^{-1} \tau_n^{-1} \sum_s h_s\left(\mathbb{W}_s - 1\right)$$

$$= n^{1/2} N^{-1} \tau_n^{-1} \sum_s \sum_{j=1}^{m} h_1(Y_{i_j})\left(\mathbb{W}_s - 1\right)$$

$$+ n^{1/2} N^{-1} \theta \tau_n^{-1} \sum_s \left(\mathbb{W}_s - 1\right) + n^{1/2} N^{-1} \tau_n^{-1} \sum_s g_s\left(\mathbb{W}_s - 1\right)$$

$$= T_1 + T_2 + T_3 \text{ say.}$$

Because of the condition $\sum_{i=1}^{n} \mathbb{W}_{n:i} = n$ we have that $\sum_s\left(\mathbb{W}_s - 1\right) = 0$, so $T_2 = 0$ almost surely.

Using (4.13) and the fact that $\mathbb{E}\big(N^{-1}\sum_s g_s\big)^4 = O(n^{-4})$ (see Serfling (1980), page 188)) and (4.9), after some algebra we have that for any $\delta > 0$,

$$\mathbb{P}_B\Big[|T_3| > \delta\Big] = O_P\big(n^{-2}\big).$$

It remains to consider T_1. For this term also, using (4.6), we have that

$$T_1 = n^{-1/2}\sum_{i=1}^n W_i h_1\big(Y_i\big) + r_{nb},$$

where again

$$\mathbb{P}_B\Big[|r_{nb}| > \delta\Big] = o_P\big(n^{-2}\big) \quad \text{for any} \quad \delta > 0.$$

We now need to show that the distribution of $n^{1/2}(U_n - \theta)$ and the bootstrap distribution of $n^{-1/2}m\sum_{i=1}^n W_i h_1(Y_i)$ converge to the same limiting normal distribution. For the original U-statistic, this is the UCLT, Theorem 1.1. For the bootstrap statistic, we use Theorem 4.4 to get the result. The conditions (4.10) and (4.11) are required in order to satisfy the conditions of Theorem 4.4. The details are left as an exercise. $\qquad\square$

4.4.1   Computational aspects for additive weights

U-statistics are computationally lengthy, since $\binom{n}{m}$ quantities are involved. Also, the computation of $h(\cdot)$ may be difficult in many cases. Thus generalized bootstrap by using (4.2) make heavy demand on computational time and storage space requirements. We address this issue now.

Define

$$\tilde{U}_{ni} = \binom{n-1}{m-1}^{-1} \sum_{1\le i_1\cdots<i_{m-1}\le n; i_j\neq i} h(Y_i, Y_{i_1},\ldots,Y_{i_{m-1}}).$$

A simple calculation shows that our generalized bootstrap statistic using *additive weights* (4.5) is of the form

$$U_{nb} = n^{-1}\sum_{i=1}^n \mathbb{W}_i \tilde{U}_{ni}. \tag{4.18}$$

In order to discuss the computational advantages of using additive resampling weights over other forms of resampling weights, we use two considerations, the computational time requirement and the storage space requirement for bootstrapping. Since Monte Carlo simulations are an integral part of the computation of bootstrap, we assume that B bootstrap iterations are carried out under both the methods.

Theorem 4.5. (a) *The time complexity of estimating the distribution of U_n using additive resampling weights and generic resampling weights in B bootstrap iterations are respectively $O(n(B+1) + n^m)$ and $O(n^m(B+1))$ respectively.*

(b) *The space complexity of estimating the distribution of U_n using additive resampling weights and generic resampling weights are $O(n)$ and $O(n^m)$ respectively, in addition to the space requirement for storing the results of each bootstrap iteration.*

The remarkable fact is that while using additive weights (4.5), the time and storage space requirements are reduced simultaneously, and for each bootstrap step instead of requiring $O(n^m)$ time and space, the requirement is only $O(n)$.

Proof of Theorem 4.5: (a) Let us assume that for given $(y_1, \ldots, y_n)$ the computation of each $h(y_1, \ldots, y_m)$ takes H units of time. Then it can be seen that each $\tilde{U}_{ni}$ requires $\left(\binom{n-1}{m-1}\right) H$ units of time, and once all $\tilde{U}_{ni}$'s have been computed and stored, U_n is easily computed in n steps. Thus an initial n^m order computation is required for both the bootstrap methods.

However, once the bootstrap weights are generated (without loss we assume these to be generated in $O(n)$ time), for the additive weights case only $2n$ more steps are needed, whereas for the general resampling weights case the number of steps needed are $m\binom{n}{m}$ assuming all the $h(Y_{i_1}, \ldots, Y_{i_m})$ are stored. Thus for each bootstrap iteration the time complexities are $O(n)$ and $O(n^m)$ respectively for the two methods. This completes the proof of part (a).

Part (b) is easily proved by observing that when we use additive weights only the $\tilde{U}_{ni}$ defined in (4.18) need to be stored, whereas when we use general resampling weights all the $h(Y_{i_1}, \ldots, Y_{i_m})$ have to be stored. $\qquad\square$

4.5 Generalized bootstrap for M_m-estimators

We now focus on estimating the distribution of an M_m-estimator by resampling. Since an M_m-estimator is obtained by a minimisation procedure, it is natural to obtain its bootstrap version by a parallel minimisation which involves random weights. This is precisely how several resampling schemes were seen to operate in linear regression in Chapter 3.

We first establish bootstrap asymptotic representation results, following which we obtain consistency results for the generalized bootstrap of M_m-estimators. We recall the setup from Chapter 2.

Let $Y_1, \ldots, Y_m$ be m i.i.d. copies of a $\mathbf{Y} \in \mathcal{Y} \subseteq \mathbb{R}^p$ valued random variable and let $f(y, a)$ be a real measurable function defined for $y \in \mathcal{Y}^m$ and $a \in \mathbb{R}^d$. Let

$$Q(\theta) = \mathbb{E}f(Y_1, \ldots, Y_m; \theta). \qquad (4.19)$$

Let $\theta_0 \in \mathbb{R}^d$ be such that

$$Q(\theta_0) = \min_a Q(a). \qquad (4.20)$$

We assume θ_0 is unique, and it is the unknown parameter, to be estimated from the data.

Suppose that $Y_1, \ldots, Y_n$ is an i.i.d. sample. Then the sample analog of (4.19) is

$$Q_n(\theta) = \binom{n}{m}^{-1} \sum_{1 \le i_1 < \ldots i_m \le n} f(Y_{i_1}, Y_{i_2}, \ldots, Y_{i_m}; \theta). \qquad (4.21)$$

Let $\hat{\theta}_n$ be any measurable selection such that

$$Q(\hat{\theta}_n) = \min_a Q_n(a), \qquad (4.22)$$

which is the M_m-estimator of the parameter θ_0.

As in Chapter 2, we assume that $f(y; a)$ is convex in a. Let $g(y; a)$ be a measurable subgradient of $f(y; a)$, that is

$$f(y; a) + (b - a)^T g(y; a) \le f(y; b) \qquad (4.23)$$

holds for all $a, b \in \mathbb{R}^d$, $y \in \mathcal{Y}^m$. Recall the assumptions needed in Chapter 2 for the asymptotic normality of M_m estimators.

(I) $f(y; a)$ is convex with respect to a for each fixed y.

(II) (4.19) exists and is finite for all θ, that is, $Q(\theta)$ is well defined.

(III) θ_0 satisfying (4.20) exists and is unique.

(IV) $\mathbb{E}|g(Y_1, \ldots, Y_m; a)|^2 < \infty$ for all a in a neighborhood of θ_0.

(V) $Q(a)$ is twice differentiable at θ_0 and $H = \nabla^2(\theta_0)$ is positive definite.

The bootstrap equivalent of $\hat{\theta}_n$ is obtained by minimizing

$$Q_{nb}(\theta) = \binom{n}{m}^{-1} \sum_{1 \le i_1 < \ldots i_m \le n} \mathbb{W}_{n:i_1,i_2,\ldots,i_m} f(Y_{i_1}, Y_{i_2}, \ldots, Y_{i_m}; \theta). \quad (4.24)$$

Let $\{\hat{\theta}_{nB}\}$ be chosen in a measurable way (such a choice is possible) satisfying

$$Q_{nb}(\hat{\theta}_{nb}) = \min_{\theta} Q_{nb}(\theta). \quad (4.25)$$

This $\hat{\theta}_{nb}$ is not unique in general.

4.5.1 Resampling representation results for $m = 1$

In this subsection we consider the $m = 1$ case separately for easier understanding of the general case that we present later.

Theorem 4.6 (Bose and Chatterjee (2003)). *Suppose Assumptions (I)-(V) hold. Assume that resampling is performed by minimizing (4.24) with weights that satisfy (4.7)-(4.9). Then*

$$\tau_n^{-1} n^{1/2}(\hat{\theta}_{nb} - \hat{\theta}_n) = -n^{-1/2} H^{-1} S_{nb} + r_{nb} \ \text{where} \quad (4.26)$$

$$S_{nb} = \sum_{i=1}^{n} W_i g(Y_i; \theta_0) \quad \text{and} \quad (4.27)$$

$$\mathbb{P}_B[||r_{nb}|| > \epsilon] = o_P(1) \ \text{for any} \ \epsilon > 0. \quad (4.28)$$

Proof of Theorem 4.6: In order to prove this theorem, we use the triangulation Lemma of Niemiro (1992) which is also given in Chapter 2. Without loss of generality, let $\theta_0 = 0$ and $Q(\theta_0) = 0$. Now define

$$X_{ni} = [f(Y_i; n^{-1/2}\theta) - f(Y_i; 0)] - n^{-1/2}\theta^T g(Y_i; 0) \quad \text{and}$$
$$X_{nbi} = \mathbb{W}_i X_{ni}.$$

Then we have

$$\mathbb{E}_B X_{nbi} = X_{ni} \text{ and}$$

$$\sum_{i=1}^{n} X_{nbi} = n Q_{nb}(n^{-1/2}\theta) - n Q_{nb}(0) - n^{-1/2}\theta^T S_{nb} \text{ where}$$

$$S_{nb} = \sum_{i=1}^{n} \mathbb{W}_i g(Y_i; 0).$$

In order to make explicit the fact that X_{ni} is a function of θ, we sometimes write $X_{ni}(\theta)$.

Fix $\delta_1, \delta_2 > 0$. We first prove two facts.

(a) For any $M > 0$

$$\mathbb{P}_B[\sup_{||\theta|| \le M} \tau_n^{-1} |\sum X_{nbi} - \theta^T H\theta/2| > \delta_2] = o_P(1). \qquad (4.29)$$

(b) There exists $M > 0$ such that

$$\mathbb{P}[\mathbb{P}_B[||n^{-1/2} S_{nb}|| \ge M] > \delta_1] < \delta_2. \qquad (4.30)$$

Proof of (4.29). Fix an $M > 0$. For fixed $\delta_* > 0$, get $\delta > 0$ and $\epsilon > 0$ such that $M_1 = M + (2\delta)^{1/2}$ and $\delta_* > 5M_1\lambda_{max}(H)\delta + 3\epsilon$, where $\lambda_{max}(H)$ is the maximum eigenvalue of H. Consider the set $A_1 = \{\theta : ||\theta|| \le M_1\}$ and let $B = \{b_1, \ldots, b_N\}$ be a finite δ-triangulation of A_1. Note that with

$$A = \{\theta : ||\theta|| \le M\}, \quad L = M_1\lambda_{max}(H),$$

$$h(\theta) = \theta^T H\theta/2, \text{ and } h'(\theta) = \sum \mathbb{W}_i X_{ni}(\theta),$$

all the conditions of Lemma 2.3 are satisfied. Now

$$\mathbb{P}_B[\sup_{||\theta|| \le M} \tau_n^{-1} |\sum X_{nbi} - \theta^T H\theta/2| > \delta_*]$$

$$\le \mathbb{P}_B[\sup_{||\theta|| \le M} \tau_n^{-1} |\sum X_{nbi} - \theta^T H\theta/2| > 5L\delta + 3\epsilon]$$

$$\leq \mathbb{P}_B[\sup_{\theta \in B} \tau_n^{-1} | \sum X_{nbi} - \theta^T H\theta/2| > \epsilon] \tag{4.31}$$

$$\leq \sum_{j=1}^{N} \mathbb{P}_B[\tau_n^{-1} | \sum \mathbb{W}_i X_{ni}(b_j) - b_j^T H b_j/2| > \epsilon]$$

$$\leq \sum_{j=1}^{N} \mathbb{P}_B[|\sum W_i X_{ni}(b_j)| > \epsilon/2] + \sum_{j=1}^{N} \mathcal{I}_{\{\tau_n^{-1}|\sum X_{ni}(b_j) - b_j^T H b_j/2| > \epsilon/2\}}$$

$$\leq \sum_{j=1}^{N} k \sum_{i=1}^{n} X_{ni}^2(b_j) + \sum_{j=1}^{N} \mathcal{I}_{\{\tau_n^{-1}|\sum X_{ni}(b_j) - b_j^T H b_j/2| > \epsilon/2\}} \tag{4.32}$$

$$= o_P(1). \tag{4.33}$$

In the above calculations (4.31) follows from the triangulation Lemma 2.2 given in Chapter 2. Observe that in (4.32) the index j runs over finitely many points, so it is enough to show the probability rate (4.33) for each fixed b. It has been proved by Niemiro (1992) (see page 1522) that for fixed b,

$$\sum_i X_{ni}^2(b) = o_P(1) \quad \text{and} \quad \sum_i X_{ni}(b) - b^T H b/2 = o_P(1). \tag{4.34}$$

Hence (4.33) follows by using the lower bound in (4.8). This proves (4.29).
Proof of (4.30). Note that

$$\mathbb{P}_B[||n^{-1/2} S_{nb}|| > M] \leq \frac{1}{M^2 n} \mathbb{E}_B || \sum_{i=1}^{n} \mathbb{W}_i g(Y_i; 0)||^2$$

$$\leq \frac{2}{M^2 n} [\tau_n^{-2} \mathbb{E}_B || \sum_{i=1}^{n} W_i g(Y_i; 0)||^2 + || \sum_{i=1}^{n} g(Y_i; 0)||^2]$$

$$\leq \frac{K \tau_n^{-2}}{M^2 n} \sum_{i=1}^{n} ||g(Z_i; 0)||^2 + \frac{2}{M^2 n} || \sum_{i=1}^{n} g(Y_i; 0)||^2$$

$$= U_M \text{ say.}$$

The constant K in the last step is obtained by using the upper bound condition in (4.8) and (4.9). Now fix any two constants $\epsilon, \delta > 0$. By choosing M large enough, we have

$$\mathbb{P}[\mathbb{P}_B[\tau_n^{-1}||n^{-1/2} S_{nb}|| < M] > \epsilon]$$
$$\leq \mathbb{P}[\epsilon < U_M]$$
$$< \delta \tag{4.35}$$

by using (IV). This proves (4.30).

Fix a $\delta_0 > 0$. Now on the set $\mathcal{A}$ where both

$$\sup_{||\theta|| \le M} \tau_n^{-1} || \sum_{i=1}^{n} X_{nbi} - \frac{1}{2} \theta^T H \theta || < \delta_0, \tag{4.36}$$

$$||n^{-1/2} H^{-1} S_{nb}|| < M - 1 \tag{4.37}$$

hold, the convex function $nQ_{nb}(n^{-1/2}\theta) - nQ_{nb}(0)$ assumes at $n^{-1/2}H^{-1}S_{nb}$ a value less than its values on the sphere

$$||\theta + n^{-1/2} H^{-1} S_{nb}|| = \kappa \tau_n \delta_0^{1/2},$$

where $\kappa = 2\lambda_{min}^{1/2}(H)$. The global minimiser of this function is $n^{1/2}\hat{\theta}_{nb}$. Hence

$$n^{1/2}\tau_n^{-1}\hat{\theta}_{nb} = -n^{-1/2}\tau_n^{-1}H^{-1}S_{nb} + r_{nb3}. \tag{4.38}$$

Since δ_0 is arbitrary, and because of (4.29) and (4.30), we have for any $\delta > 0$ $\mathbb{P}_B[||r_{nb3}|| > \delta] = o_P(1)$.

Now use Theorem 2.3 from Chapter 2 to get the result. This step again uses (4.8).   □

For any $c \in \mathbb{R}^d$ with $||c|| = 1$, let

$$\mathcal{L}_n(x) = \mathbb{P}[n^{1/2}c^T(\hat{\theta}_n - \theta_0) \le x] \quad \text{and}$$
$$\mathcal{L}_{nb}(x) = \mathbb{P}_B[\tau_n^{-1}n^{1/2}c^T(\hat{\theta}_{nb} - \theta_n) \le x].$$

The following corollary establishes the consistency of the resampling technique for estimating $\mathcal{L}_n$.

Corollary 4.1. (Consistency of resampling distribution estimate) *Assume the conditions of Theorem 4.6. Assume also (4.10)-(4.11). Then*

$$\sup_{x \in \mathbb{R}} |\mathcal{L}_{nb}(x) - \mathcal{L}_n(x)| = o_P(1) \quad \text{as } n \to \infty.$$

Proof of Corollary 4.1: Again, without loss of generality assume, $\theta_0 = 0$ and $Q(\theta_0) = 0$. Let

$$h(Y_i; 0) = c^T H^{-1} g(Y_i; 0) \quad \text{and}$$
$$V_\infty = Eh^2(Y_1; 0).$$

Using Theorem 2.3 we obtain that

$$\sup_x |\mathcal{L}_n(x) - N(0, V_\infty)| \to 0. \tag{4.39}$$

Also observe that given the condition on the bootstrap weights, Theorem 4.4 may be applied to obtain

$$\sup_x |\mathcal{L}_{nb}(x) - N(0, V_\infty)| = o_P(1) \quad \text{as} \quad n \to \infty. \tag{4.40}$$

Then (4.39) and (4.40) complete the proof. $\qquad\qquad\qquad\square$

4.5.2 Results for general m

For convenience, let us fix the notation

$$S = \{(i_1, \ldots, i_m) : 1 \le i_1 < i_2 < \cdots < i_m \le n\},$$

and a typical element of S is $s = (i_1, \ldots, i_m)$. We use the notation $|s \cap t| = k$ to denote two typical subsets $s = \{i_1, \ldots, i_m\}$ and $t = \{j_1, \ldots, j_m\}$ of size m from $\{1, \ldots, n\}$, which have exactly k elements in common.

We often use the same notation s for both $s = (i_1, \ldots, i_m) \in S$ and $s = \{i_1, \ldots, i_m\} \subset \{1, \ldots n\}$. Also, let $N = \binom{n}{m}$. Let $Y_s = (Y_{i_1}, \ldots, Y_{i_m})$ when $s = (i_1, \ldots, i_m) \in S$. The notation $\sum_{|s \cap t| = j}$ denotes sum over all s and t for which $|s \cap t| = j$ holds.

The bootstrap weights W_s are non-negative random variables. We assume that for all $s \in S$, the distribution of W_s is the same with $\mathbb{E} W_s = 1$. Define

$$\xi_n^2 = \mathbb{E}_B (W_s - 1)^2 \quad \text{and} \quad W_s = \xi_n^{-1}(W_s - 1).$$

Assume that $\mathbb{E}_B W_s W_t$ is a function of $|s \cap t|$ only, and let

$$c_j = \mathbb{E}_B W_s W_t \quad \text{whenever} \quad |s \cap t| = j, \quad \text{for} \quad j = 0, 1, \ldots, m.$$

We assume that the following conditions are satisfied:

$$K > \xi_n^2 > k > 0 \tag{4.41}$$

$$c_0 = o(1) \tag{4.42}$$

$$|c_j| = O(1), \ j = 1, \ldots, m. \tag{4.43}$$

Define

$$g_1(Y_1; \theta_0) = E[g(Y_1, \ldots, Y_m; \theta_0)|Y_1] \quad \text{and}$$

$$f_i = \binom{n-1}{m-1}^{-1} \sum_{s:i \in s} W_s,$$

where the sum runs over all $s \in S$ such that $i \in \{i_1, \ldots, i_m\}$.

Theorem 4.7 (Bose and Chatterjee (2003)). *Suppose Assumptions (I)-(V) hold. Assume that resampling is performed by minimizing (4.24) with weights that satisfy (4.41)-(4.43). Then*

$$\xi_n^{-1} n^{1/2} (\hat{\theta}_{nb} - \hat{\theta}_n) = -n^{1/2} H^{-1} S_{nb} + r_{nb} \tag{4.44}$$

$$= -mn^{-1/2} \sum f_i H^{-1} g_1(Y_i; \theta_0) + R_{nb} \tag{4.45}$$

where $g_1(Y_i; \theta_0)$ is the first projection of $g(\cdot; \theta_0)$,

$$S_{nb} = \binom{n}{m}^{-1} \sum_{s \in S} \mathbb{W}_s g(Y_s, \theta_0),$$

and

$$\mathbb{P}_B[||r_{nb}|| > \epsilon] = o_P(1) \quad \text{and} \tag{4.46}$$

$$\mathbb{P}_B[||R_{nb}|| > \epsilon] = o_P(1) \quad \text{for every } \epsilon > 0. \tag{4.47}$$

From (4.42) we have that the asymptotic mean of f_1 is 0. Let its asymptotic variance be $v^2(m)$. This will in general be a function of m. Then the appropriate standardized bootstrap statistic is

$$n^{1/2} \xi_n^{-1} v_m^{-1} (\hat{\theta}_{nB} - \hat{\theta}_n).$$

For any $c \in \mathbb{R}^d$ with $||c|| = 1$, let

$$\mathcal{L}_n(x) = \mathbb{P}[n^{1/2} c^T (\hat{\theta}_n - \theta_0) \leq x] \quad \text{and}$$

$$\mathcal{L}_{nb}(x) = \mathbb{P}_B[n^{1/2} \xi_n^{-1} v_m^{-1} c^T (\hat{\theta}_{nb} - \hat{\theta}_n) \leq x].$$

Then we have the following Corollary. Its proof is similar to the proof of the corresponding results in Section 4.5.1, and we omit the details.

Corollary 4.2. *Assume the conditions of Theorem 4.7. Assume also that $f_{n:i}$'s are exchangeable with $\sup_n \mathbb{E}_B f_{n:i}^4 < \infty$ and $\mathbb{E}_B f_{n:i}^2 f_{n:j}^2 \to 1$ for $i \neq j$ as $n \to \infty$. Then*

$$\sup_{x \in \mathbb{R}} |\mathcal{L}_{nb}(x) - \mathcal{L}(x)| = o_P(1) \quad as \ n \to \infty.$$

Similar to the development presented in Section 4.5.1, the representation (4.45) will turn out to be useful for showing consistency of the generalized bootstrap scheme, both for variance estimation as well as distributional approximation.

Proof of Theorem 4.7: The first part of this proof is similar to the proof of Theorem 4.6. Define

$$X_{ns} = f(Y_s; n^{-1/2}\theta) - f(Y_s; 0) - n^{-1/2}\theta^T g(Y_s; 0),$$
$$X_{nbs} = W_s X_{ns}, \quad \text{and} \quad S_{nb} = N^{-1} \sum_s g(Y_s; 0).$$

We have

$$nQ_{nb}(n^{-1/2}\theta) - nQ_{nb}(0) - n^{1/2}\theta^T S_{nb} - \theta^T H\theta/2$$
$$= nN^{-1} \sum X_{nbs} - \theta^T H\theta/2.$$

Then arguments similar to those used in the proof of Theorem 4.6 yields (4.44) once it is established that for any fixed $\epsilon, \delta > 0$,
(a) For any $M > 0$,

$$\mathbb{P}_B[\sup_{||\theta|| \leq M} |nN^{-1} \sum X_{nbs} - \theta^T H\theta/2| > \delta] = o_P(1). \tag{4.48}$$

(b) There exists $M > 0$ such that,

$$\mathbb{P}[\mathbb{P}_B[||n^{1/2}S_{nb}|| > M] > \epsilon] < \delta. \tag{4.49}$$

The proof of these are similar to those of (4.29) and (4.30). By carefully following the arguments for those, we only have to show

$$n^2 N^{-2} \mathbb{E}_B (\sum W_s X_{ns})^2 = o_P(1), \tag{4.50}$$
$$nN^{-2} \mathbb{E}_B || \sum w_s g(Y_s, 0)||^2 = O_P(1). \tag{4.51}$$

For (4.50), we have

$$n^2 N^{-2} \mathbb{E}_B \left(\sum W_s X_{ns} \right)^2$$

$$= n^2 N^{-2} \sum_{j=0}^m c_j \sum_{|s \cap t| = j} X_{ns} X_{nt}$$

$$= n^2 N^{-2} c_0 \left(\sum_s X_{ns} \right)^2 + n^2 N^{-2} \sum_{j=1}^m c_j \sum_{|s \cap t| = j} X_{ns} X_{nt}.$$

A little algebra shows that $nN^{-1} \sum X_{ns} = O_P(1)$. The first term is $o_P(1)$ from this and (4.42). For the other term, first note that the sum over j is finite, and from (4.43), we only need to show

$$n^2 N^{-2} \sum_{|s \cap t| = j} X_{ns} X_{nt} = o_P(1) \quad \text{for every fixed} \quad j = 1, \dots, m.$$

For this, we have

$$n^2 N^{-2} \sum_{|s \cap t| = j} X_{ns} X_{nt}$$

$$\leq n N^{-2} \sum_{|s \cap t| = j} [\theta^T (g(Y_s, n^{-1/2}\theta) - g(Y_s, 0))][\theta^T (g(Y_t, n^{-1/2}\theta) - g(Y_t, 0))].$$

Since the number of terms in $\sum_{|s \cap t| = j}$ is $O(n^{-1} N^2)$, it is enough to show that for any $s, t \in S$

$$a^T (g(Y_s, n^{-1/2}\theta) - g(Y_s, 0)) \theta^T (g(Y_t, n^{-1/2}\theta) - g(Y_t, 0)) = o_P(1).$$

Observe that $\theta^T (g(Y_s, n^{-1/2}\theta) - g(Y_s, 0))$ are non-negative random variables that are non-increasing in n, and their limit is 0. This establishes the above. Details of this argument is similar to those given in Chapter 2 following (2.33). The proof of (4.51) is along the same lines. This proves (4.44).

In order to get the representation (4.45), we have to show

$$n^{1/2} N^{-1} \sum W_s g(Y_s, 0) = m n^{-1/2} \sum_i f_i g_1(Y_i, 0) + R_{nb1},$$

where

$$\mathbb{P}_B [\|R_{nb1}\| > \delta] = o_P(1) \quad \text{for any} \quad \delta > 0.$$

Let $h(Y_s, 0) = g(Y_s, 0) - \sum_{j=1}^m g_1(Y_{i_j}, 0)$. This is a kernel of a first order degenerate U-statistic. Then we have

$$W_s g(Y_s, 0) = W_s \sum_{j=1}^m g_1(Y_{i_j}, 0) + W_s h(Z_s, 0).$$

Also, it can be established that

$$\sum_s \sum_{j=1}^m W_s g_1(Y_{i_j}; 0) = \binom{n-1}{m-1} \sum_{i=1}^n f_i g_1(Y_i; 0).$$

Also

$$\mathbb{E}(\|N^{-1} \sum h(Y_s, 0)\|)^{-2} = O(n^{-2}). \tag{4.52}$$

Now using this result and (4.43), after some algebra we obtain

$$\mathbb{P}_B[\|n^{1/2} N^{-1} \sum W_s h(Z_s, 0)\| > \delta] = o_P(1),$$

and this yields (4.45). □

For different choices of generalized bootstrap weights, and defining $\mathbb{W}_s$ either as in (4.4) or (4.5), the conditions (4.41)-(4.43) are satisfied. When using (4.4), the moments of products of $\mathbb{W}_i$'s of various orders are involved and conditions are difficult to check unless m is small.

A natural question is how f_i and W_s relate to each other. Observe that in both the multiplicative and additive form of $\mathbb{W}_s$ given in (4.4) and (4.5), the difference between W_s and an appropriately scaled f_i (where the scaling factor is a function of m only) is negligible in the representation (4.45). Again the algebra is complicated in case of (4.4), and may be verified for small values of m. There are no such difficulties with (4.5).

4.6 Exercises

1. Suppose $w_1, \ldots w_k$ are exchangeable random variables such that $\sum_{i=1}^k w_i$ is a constant. Show that for every $m \neq n$, $c_{11} = corr(w_m, w_n) = -\frac{1}{k-1}$ irrespective of the distribution of $\{w_i\}$. Hence verify (4.12).

2. Suppose $e_i, 1 \leq i \leq k$ are k-dimensional vectors where the ith position of e_i is 0 and the rest of the positions are 1. Consider the k-variate distribution which puts mass $1/k$ at each of these vectors. That is, consider the random vector w such that $P(w = e_i) = 1/k, 1 \leq i \leq k$. Let w_i be the ith coordinate of w. Show that

   (i) $\{w_i\}$ are exchangeable, and each w_i has a Bernoulli distribution.

   (ii) for any $n \neq m$, $Corr(w_n, w_m) = -\frac{1}{k-1}$.

   (iii) as $k \to \infty$, for any fixed i, w_i has a degenerate distribution.

3. Extend the setup in the previous question to the case where each possible k-dimensional vectors has d coordinates 0 and rest have 1's.

4. Suppose $w_i, 1 \leq i \leq k$ has a multinomial distribution with number of trials k and equal success probability $1/k$ for each cell. Show that

   (i) $\{w_i\}$ are exchangeable, and each w_i has a Bernoulli distribution.

   (ii) for any $n \neq m$, $Corr(w_n, w_m) = -\frac{1}{k-1}$.

   (iii) as $k \to \infty$, for any fixed i, the asymptotic distribution of w_i is Poisson with mean 1.

   (iv) as $k \to \infty$, for any fixed $i \neq j$, w_i and w_j are asymptotically independent.

5. Suppose X_i are i.i.d. with mean μ and variance σ^2. Suppose $w_{n,i}$ are multinomial with number of trials n and equal cell probabilities $1/n$. Show that as $n \to \infty$, the distribution of $n^{-1/2} \sum_{i=1}^n (X_i w_{n,i} - X_i)$, conditional on $X_1, \ldots X_n, \ldots$, converges almost surely to the normal with mean zero and variance σ^2. Show that the distribution of $n^{-1/2} \sum_{i=1}^n (X_i - \mu)$ converges to the same asymptotic limit.

6. Justify how the minimiser in (4.25) can be chosen in a measurable way.

7. Prove the following fact used in the proof of Theorem 4.3:

$$\mathbb{E}\left(N^{-1} \sum_s g_s\right)^4 = O(n^{-4}).$$

8. Prove the following fact used in the proof of Theorem 4.3:
   $n^{-1/2} m \sum_{i=1}^n W_i h_1(Y_i)$, conditionally on the data $\{Y_1, \ldots, Y_n, \ldots\}$ has an asymptotic $N(0, \sigma_1^2)$ distribution where $\sigma_1^2 = \mathbb{V}h_1(Y_i)$.

9. Prove (4.34) used in the proof of Theorem 4.6.

10. Show that (4.40) follows from an application of Theorem 4.4.

11. Prove Corollary 4.2.

Chapter 5

An Introduction to R

5.1 Introduction, installation, basics

This chapter is a soft introduction to the statistical software called R. We will discuss how to conduct elementary data analysis, use built-in programs and packages, write and run one's own programs, in the context of the topics covered in this book. All softwares have some specific advantages and several deficiencies, and R is no exception. The two main reasons we consider R as the software of choice are that it is **completely free**, and that this is perhaps the most popular software among statisticians. Because of these reasons, a very large number of statistical algorithms and procedures have been already developed for the R environment. We have written a software package called *UStatBookABSC* to accompany this book, containing as built-in functions the R codes for several of the procedures discussed here. We discuss this particular R package later in this chapter.

There are many texts and online notes on R, some are introductory, while others present the concepts from statistical and computational viewpoints at various levels of detail. A very small sample of recent references is Dalgaard (2008); Matloff (2011). Many more are available for free download from the internet, or for purchase.

For any questions relating to R, generally internet searches are adequate to elicit answers or hints on how to proceed. This software runs on all standard operating systems, and the example codes that we present below will run on various versions of Linux, Unix, Apple Macintosh™ and Microsoft

© Springer Nature Singapore Pte Ltd. 2018 and Hindustan Book Agency 2018
A. Bose and S. Chatterjee, *U-Statistics, Mm-Estimators and Resampling*, Texts and Readings in Mathematics 75, https://doi.org/10.1007/978-981-13-2248-8_5

Windows^{TM}, with occasional minor modifications. We have strived for simplicity and transparency and most of the development below is strictly for pedagogical purposes. Some of the displayed output is edited in a minor way to fit on the page. We assume essentially no knowledge of programming languages or computational software.

The homepage for R is `https://www.r-project.org/`. Here, there are instructions and links for downloading and installing the software. The first step is to download and install the `base` distribution. Once installation of the `base` distribution is complete and successful, there ought to be an icon on the desktop (and/or elsewhere) that looks like the English letter R, generally accompanied with some numbers, denoting the version number, in small print below it.

Click on the R icon, and an R workspace comes up. The major component of this is the `R console`, where one can type in data, R commands, and where the standard on-screen output of R is shown. Graphics, like histograms, boxplots or scatter plots, show up in a different panel.

Note the short description of R that comes right at the top of the `R console`. Note in particular, that the software comes without any warranty. A second statement says that R is a project with many contributors; many such contributors have provided macro-programs to do specific tasks in R. Such "contributed packages" are extremely useful, and later in this chapter we discuss how to download, install, update and use them. Note however, that R is *open source*, and while there are several extremely useful and important contributed packages, there is no quality guarantee of contributed packages. The third statement in the `R console` is about how to use the `help`, `demo`, and how to quit R. Many users install and use R-based *integrated development environment* (IDE) software applications like `R Studio`, available from `https://www.rstudio.com/`. Such IDE are often favored by many users, and have several convenient features.

R is comprised of the core, `base` package, and several additional contributed packages. The `base` package is necessary. Once that is installed and running, one may download and install any of the contributed packages as and when needed. The package accompanying this book, `UStatBookABSC`, is one such contributed package. The simplest way of installing a contribution package is to click on `packages` in the top-menubar in the R workspace. Other options, like using the command `install.packages` at the R workspace prompt, or using the command line `R CMD INSTALL` are also available.

Once a package, say `UStatBookABSC` (Chatterjee (2016)) is installed in one's computer using any of the above methods, the following command may be used to access its contents:

```
> library(UStatBookABSC)
```

Packages and core R may occasionally need to be updated, for which the steps are mostly similar to installation. Package installation, updating and new package creation can also be done from inside R Studio, in some ways more easily.

As a general rule, using the R help pages is very highly recommended. They contain a lot more information than what is given below. The way to obtain information about any command, say the *print* command, is to simply type in

```
> ?print
```

If the exact R command is not known, just searching for it in generic terms on the internet usually elicits what is needed.

Simple computations can be done by typing in the data and commands at the R prompt, that is the ">" symbol in the R console. However, it is not good practice to type in lengthy commands or programs in the R console prompt, or corresponding places in any IDE. One should write R programs using a text editor, save them as files, and then run them. R programs are often called R scripts.

One common way of writing such *scripts* or programs is by using the built-in editor in R. In the menubar in the R workspace, clicking on the File menu followed by the New script menu opens the editor on an empty page where one may write a new program. To edit an existing script, the menu button Open script may be used. In order to run a program/script, one should click on the Source menu button. The Change dir menu button allows one to switch working directory.

To check the working directory, use the command `getwd()`.

```
> getwd()
[1] "/Users/ABSC/Programs"
```

To change the working directory we use the command `setwd()`;.

```
> setwd("/Users/ABSC/Programs/UStatBook")
> getwd()
[1] "/Users/ABSC/Programs/UStatBook"
```

Sometimes, but not always, one might want to come back to a work done earlier on an R workspace. To facilitate this, R can save a workspace at the end of a session. Also, when starting a new R session, one may begin with a previously saved workspace. All of these are useful, *if and when* we want to return to a previous piece of work in R. It is an inconvenience, however, when one wants to do fresh work, and does not want older variable name assignments and other objects stored in R memory to crop up. Also, a saved workspace often takes up considerable storage space. Memory problems are also known to occur. These issues are not huge problems for experts, but often inconvenience beginners.

5.1.1   Conventions and rules

Note that R is case sensitive, thus `a` and `A` are considered different. In the rest of the chapter, note the use of upper and lowercase letters throughout the R commands. In particular, many standard built-in R commands like `help, print, plot, sum, sqrt, mean, var, cov, cor, median` are in lowercase, while others like `R CMD BATCH` are in uppercase, and some like `Rscript` or `UStatBookABSC` have both.

The symbol `#` in an R script denotes a comment, which is skipped by R when executing the script. Thus, this is an extremely convenient tool to annotate and write remarks, explanatory notes and comments in R programs.

It is a good practice to liberally annotate and provide explanatory notes in R programs.

As with many other computer programming languages, the command x = y in R means the following: *the value of the object y is stored in the space that we will denote by the name x.* For example, x = 5 means we are storing the number 5 in a particular chamber of the computer's memory, that we will call x, and that we can recall, edit, change or use, simply by it's name x. Instead of just assigning a number to the name/memory-space x, we might assign a vector, matrix, data-frame, or any other "object". A popular convention, sometimes assiduously followed by R programmers, is to use x <- 5 for assigning the number 5 to the address x.

5.2 The first steps of R programming

As mentioned earlier, an effective and simpler way of programming in R is to type in the commands in a file, and then execute those commands collectively using the **Source file** command on the top menubar. For example, we may create the file UStatBookCodes.R for the computational work related to this section, and save it in the directory **/Users/ABSC/Programs/UStatBook**. This may be done either externally in a text editor, or by using the "New File" command from the R console. In order to clear out the contents that may be lodged in the memory of R including removal of previously used variables and functions, one may use the command

```
rm (list = ls())
```

as the first line of the file UStatBookCodes.R. We run this file in R by clicking on the "Source File" command. Thus

```
> source("/Users/ABSC/Programs/UStatBook/UStatBookCodes.R")
```

executes all the commands that we save in the file UStatBookCodes.R, which we have saved in the directory **/Users/ABSC/Programs/UStatBook**.

In fact, it is not necessary to start an interactive R session for running a script saved as a file. At the terminal ("shell" in Unix, "command prompt" in some other systems), the command

```
$ R CMD BATCH UStatBookCodes.R UStatBookOutFile.txt &
```

will run R to execute all the commands in the file UStatBookCodes.R, and save any resulting output to the file UStatBookOutFile.txt. The ampersand at the end is useful for Linux, Unix and Macintosh users, whereby the processes can be run in the background, and not be subject to either inadvertent stopping and will not require continuous monitoring. Being able to execute R files from the terminal is extremely useful when running large programs, which is often the case for Statistics research. A more recent alternative to the R CMD BATCH command is Rscript. Considerable additional flexibility is available for advanced users, the help pages contain relevant details.

One can work with several kinds of variables in R. Each single, or scalar, variable can be of type numeric, double, integer, logical or character. A collection of such scalars can be gathered in a vector, or a matrix. For example, the command

```
> x = vector(length = 5)
```

declares x to be a vector of length 5. If we type in

```
> print(x)
```

the output is

```
[1] FALSE FALSE FALSE FALSE FALSE
```

This shows that a vector of length 5 has been created and assigned the address x. The "*" operation between two vectors of equal length produces another vector of same length, whose elements are the coordinatewise product. Thus,

```
> z = x*y
> print(z)
 [1]   1  38 -96  32
```

One use of the above is to get inner products, as in sum (x * y):

```
> InnerProduct = sum (x * y)
> print(InnerProduct)
[1] -25
```

Similarly, we can get the norm of any vector x as follows:

```
> Norm = sqrt(sum( x * x))
> print(Norm)
[1] 38.07887
```

The functions InnerProduct and Norm above are built in functions in the package UStatBookABSC for users, with some additional details to handle the case where the vectors x and y have missing values.

5.3 Initial steps of data analysis

In this section, we discuss some simple data analysis steps that ought to be part of the initial steps of any scientific analysis. Perhaps the most important initial step after reading in the data is to get some idea about the nature of the variables, as with the str(D) command. Then, one may wish to obtain some summary statistics for the one dimensional marginal distribution of each variable, and consider two dimensional marginals to understand the relationship between the variables.

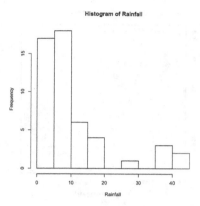

Figure 5.1: Histogram of rainfall amounts on rainy days in Kolkata during the monsoon season of 2012.

5.3.1   A dataset

For illustration purposes, we shall use the data on precipitation in Kolkata, India in 2012, during the months June to September, which corresponds approximately to the monsoon season. It consists of fifty-one rows and four columns, where the columns are on the date of precipitation, the precipitation amount in millimeters, the maximum and the minimum temperature for that day in degree Celcius. Days in which the precipitation amount was below 0.099 millimeters are not included. Suppose for the moment this dataset is available as a *comma separated value* (csv) file in our computer in the directory /Users/ABSC/Programs/UStatBook under the filename Kolkata12.csv. We can insert it in our R session as a data.frame called Kol_Precip as follows

```
> setwd("/Users/ABSC/Programs/UStatBook")
> Kol_Precip = read.csv(file = "Kolkata12.csv")
```

We urge the readers to experiment with different kinds of data files, and read the documentation corresponding to ?read.table. For example, readers may consider reading in other kinds of text files where the values are not separated by commas, where each row is not necessarily complete, and where missing values are depicted in various ways, and where the data resides in a

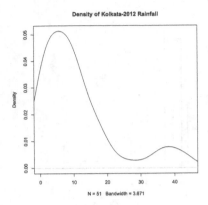

Figure 5.2: Density plot of rainfall amounts on rainy days in Kolkata during the monsoon season of 2012.

directory different from the one where the R code resides.

For advanced users, the function scan and various other functions in specialized packages are available for reading in more complex files. For example, the package fmri contains the function read.NIFTI for reading fMRI image datasets, and the package ncdf4 has functions for reading NetCDF format commonly used for storing climate and related datasets.

For convenience of the readers, we have included the above dataset on precipitation in Kolkata in two formats in the package UStatBookABSC. The first is a format which is convenient both for the reader and for R, and may be accessed as follows:

```
> library(UStatBookABSC)
>data(CCU12_Precip)
>ls(CCU12_Precip)
>?CCU12_Precip
```

The last command above brings up the help page for the dataset, which also contains an executable example.

Suppose after accessing this data inside R, we wish to save a copy of it as a comma separated value text file under the name filename Kolk12Precip.csv. This is easily done inside R as follows

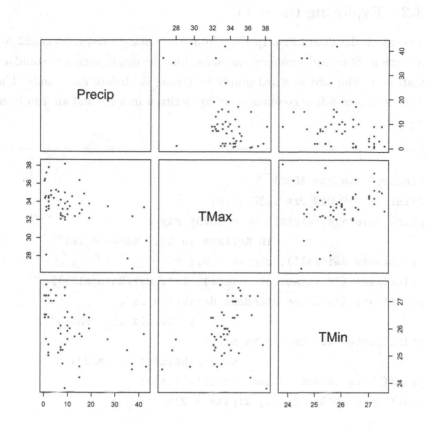

Figure 5.3: Scatter plots of rainfall, maximum and minimum temperature amounts on rainy days in Kolkata during the monsoon season of 2012.

```
> library(UStatBookABSC)
>data(CCU12_Precip)
> write.csv(CCU12_Precip, file = "Kolk12Precip.csv",
row.names = FALSE)
```

As in the case of reading of files, writing of files can also be done in various formats.

5.3.2 Exploring the data

We consider the CCU12_Precip data from the package UStatBookABSC for this section. Standard summary statistics, like the mean, variance, standard deviation, median are obtained simply by typing in obvious commands. The following easy to follow commands set are written in a file and are run from the source:

```
library(UStatBookABSC)
Rainfall = CCU12_Precip$Precip;
print("Average rainfall on a rainy day
                    in Kolkata in 2012 monsoon is:")
print(mean(Rainfall), digits = 3);
print(paste("Variance of rainfall is", var(Rainfall)));
print(paste("and the standard deviation is",
                            sd(Rainfall), "while"));
print(paste("the median rainfall is",
                        median(Rainfall), "and"));
print("here is some summary statistics");
print(summary(Rainfall), digits = 2);
```

and this produces the output

```
> print("Average rainfall on a rainy day
                    in Kolkata in 2012 monsoon is:")
[1] "Average rainfall on a rainy day
                    in Kolkata in 2012 monsoon is:"
> print(mean(Rainfall), digits = 3);
[1] 10.7
> print(paste("Variance of rainfall is", var(Rainfall)));
[1] "Variance of rainfall is 121.999325490196"
> print(paste("and the standard deviation is",
                            sd(Rainfall), "while"));
[1] "and the standard deviation is
```

```
                      11.0453304835209 while"
> print(paste("the median rainfall is",
                          median(Rainfall), "and"));
 [1] "the median rainfall is 7.9 and"
> print("here is some summary statistics");
 [1] "here is some summary statistics"
> print(summary(Rainfall), digits = 2);
   Min. 1st Qu.  Median    Mean 3rd Qu.    Max.
    0.3     2.0     7.9    10.7    14.0    42.9
```

Note that `digits` in `print` command refers to the minimum number of significant digits to print. See help manual for details and other formatting options.

The above R commands and output demonstrate how to use some built-in simple functions like `mean, var, sd, median, summary`. Note that the median value of rainfall in millimeters seem considerably less compared to the mean, which is also reflected in the gap between the third quartile of the univariate empirical distribution of `Rainfall`. More details about the univariate empirical distribution can be obtained using the `ecdf` function.

One should consider visual depictions of the data also. We run the following commands

```
hist(Rainfall);
dev.new();
plot(density(Rainfall), xlim = c(0,45),
               main = "Density of Kolkata-2012 Rainfall");
```

and obtain the histogram and the density plot of the `Rainfall` data presented respectively in Figures 5.1 and 5.2. The command `dev.new()` requires R to put the second plot in a separate window.   Note the additional details supplied to the `plot` command for the density plot, which controls the limit of the `x-axis` of the plotting region, and places the title on top of the plot. There are many more parameters that can be set to make the graphical output from R pretty, and readers should explore those. In fact, one great advantage of R as a software is its versatility in graphics. Also note that such graphical outputs can be saved automatically in several different formats.

If we run the commands

```
MaxTemp = CCU12_Precip$TMax;
MinTemp = CCU12_Precip$TMin;
print("The covariance of the max and min temperature is ");
print(cov(MaxTemp, MinTemp));
```

we get

```
> print("The covariance of the max
                        and min temperature is ");
[1] "The covariance of the max and min temperature is "
> print(cov(MaxTemp, MinTemp));
[1] 0.5833529
```

which obtains the covariance between the maximum and minimum temperature on a rainy day as 0.58. We might want to test if the difference between the maximum and minimum temperature on those days is, say, 20 degrees Celcius, and one way of conducting such a test is by using the t.test as follows:

```
> t.test(MaxTemp, MinTemp, mu = 10)
Welch Two Sample t-test
data:  MaxTemp and MinTemp
t = -7.1276, df = 67.392, p-value = 8.707e-10
alternative hypothesis: true difference
                        in means is not equal to 10
95 percent confidence interval:
 6.955590 8.287547
sample estimates:
```

```
mean of x mean of y
 33.55686   25.93529
```

Note that we do *not* recommend the above two-sample test for the present data: the maximum and minimum temperature for a given day are very likely related, and we have not verified that assumptions compatible with a two-sample t.test hold. The above computation is merely for displaying the syntax of how to conduct a two-sample test in R.

Let us now conduct a *paired t-test*, perhaps with the alternative hypothesis that the true difference is less than 10 degree Celcius keeping in mind that Kolkata is in a tropical climate region. Additionally, suppose we want a 99% one-sided confidence interval. This is implemented as follows:

```
> t.test(MaxTemp, MinTemp, mu = 10, paired = TRUE,
               alternative = "less", conf.level = 0.99)
Paired t-test
data:  MaxTemp and MinTemp
t = -7.9962, df = 50, p-value = 8.429e-11
alternative hypothesis: true difference
                          in means is less than 10
99 percent confidence interval:
    -Inf 8.336408
sample estimates:
mean of the differences
            7.621569
```

Additionally, we may decide not to rely on the t.test only, and conduct a *signed-rank test*.

```
> wilcox.test(MaxTemp, MinTemp, mu = 10, paired = TRUE,
alternative = "less", conf.int = TRUE, conf.level = 0.99)
Wilcoxon signed rank test with continuity correction
```

```
data:  MaxTemp and MinTemp
V = 82.5, p-value = 1.109e-07
alternative hypothesis: true location shift is less than 10
99 percent confidence interval:
        -Inf 7.900037
sample estimates:
(pseudo)median
        7.500018
```

The above steps of separating out the variables `Rainfall`, `MaxTemp`, `MinTemp`, and then running univariate or bivariate simplistic analysis is not recommended. In practice we should leave the data frame structure unaltered as far as possible. We demonstrate how to take only the three relevant variables from the dataset, compute summary statistics, compute the 3×3 variance-covariance matrix, followed by a scatterplot of each variable against the others as shown in Figure 5.3, then conduct a linear regression of the precipitation as a function of the maximum and minimum temperatures.

```
> KolData = subset(CCU12_Precip, select = -Date);
> summary(KolData)
      Precip            TMax             TMin
 Min.   : 0.30   Min.   :26.60   Min.   :23.80
 1st Qu.: 2.00   1st Qu.:32.65   1st Qu.:25.25
 Median : 7.90   Median :33.50   Median :25.90
 Mean   :10.72   Mean   :33.56   Mean   :25.94
 3rd Qu.:14.00   3rd Qu.:34.60   3rd Qu.:26.50
 Max.   :42.90   Max.   :38.20   Max.   :27.60
> var(KolData)
            Precip        TMax        TMin
Precip 121.999325 -12.4834510 -4.4887765
TMax   -12.483451   4.8145020  0.5833529
TMin    -4.488776   0.5833529  0.8643294
> plot(KolData, pch = "*");
```

```
> Kol.LM = lm(Precip ~ TMax + TMin, data = KolData);
> print(summary(Kol.LM));
Call:
lm(formula = Precip ~ TMax + TMin, data = KolData)
Residuals:
    Min      1Q  Median      3Q     Max
-22.648  -4.792  -1.078   3.646  30.336
Coefficients:
             Estimate Std. Error t value Pr(>|t|)
(Intercept) 179.7416    37.0017    4.858 1.31e-05 ***
TMax         -2.1385     0.6081   -3.517 0.000966 ***
TMin         -3.7500     1.4352   -2.613 0.011953 *
---
Residual standard error: 9.041 on 48 degrees of freedom
Multiple R-squared:  0.3568,Adjusted R-squared:    0.33
F-statistic: 13.31 on 2 and 48 DF,  p-value: 2.514e-05
```

5.3.3 Writing functions

In any programming language including R, it is important to know how to write functions, so as to have one set of codes that may be used repeatedly throughout the main program, to make programs more easily readable and interpretable, and to reduce the possibility of errors in repeated computations. All packages in R, including the package UStatBookABSC, is just a collection of such functions, with documentation and explanations on how to use those functions.

We illustrate how to write functions with two examples. The first shows how to compute the Euclidean inner product of two vectors x and y, which as we know is sum(x*y) in R notation. A simple function to do this computation is as follows:

```
I.Product = function( x, y){
T = sum(x*y)
```

```
return(T)
}
```

The first line says that I.Product is the name of the function, and that its arguments are x, y. The second line computes the function, and the third line returns to the main program the computed value of the function. Here is how this program may be used:

```
A = c( 1, 2, 3);
BVec = c(0, 2, -1);
I = I.Product(A, BVec);
print(I)
[1] 1
```

Note that we deliberately used the vectors A and BVec as arguments, the names of the arguments do not need to match how the function is written. Also, once the function I.Product is in the system, it can be used repeatedly. This may not seem a big deal for a simple function like I.Product, but in reality many functions are much more complex and elaborate, and their codification as standalone functions helps programming greatly. Also, even simple functions typically require checks and conditions to prevent errors and use with incompatible arguments. See, for example, the code for function InnerProduct in the UStatBookABSC package, which does the same computation as I.Product, but has checks in place to ensure that both the vectors used as arguments are numeric, they have the same length, and has additional methodological steps to handle missing values in either argument.

5.3.4   Computing multivariate medians

We now give two examples on computing multivariate medians.

Example 5.1 *(L₁-median computation)*: Recall the L_1-median from Example 2.7. Here, we show how to compute it, using built-in functions from the package UStatBookABSC. The function that we will use is L1Regression,

which is actually a function to compute multivariate median regression estimators. Details about its usage in a regression context is given in Section 5.4,
here we illustrate how the same function may be used to obtain multivariate medians. We would like to obtain the median of the three dimensional
random variable (Precip, TMax, TMin) from our data. This is done by
fitting the L_1 regression function to this vector, using only the intercept as a
covariate. The following code may be used:

```
>DataY = cbind(CCU12_Precip$Precip, CCU12_Precip$TMax,
CCU12_Precip$TMin);
>DataX = rep(1, length(CCU12_Precip$Precip));
>M.L1 = L1Regression(DataY, DataX);
> print(M.L1$BetaHat)
          [,1]      [,2]       [,3]
[1,] 8.154253 33.52507 25.93579
```

□

Example 5.2 *(Oja-median computation)***:** In this example, we illustrate how
to compute the Oja-median, introduced in Example 2.8. The function for
this is also built in the package UStatBookABSC. The following code obtains
the Oja-median for (Precip, TMax, TMin):

```
>Data.CCU = CCU12_Precip[,-1];
>M.Oja = OjaMedian(Data.CCU);
>print(M.Oja)
 Precip     TMax     TMin
 9.36444 33.65908 26.16204
```

Notice that the Oja-median and the L_1-median have similar, but not identical, estimates. In general, computation of Oja-median is more involved since
determinants of several matrices have to be computed for each optimization
step.   □

5.4   Multivariate median regression

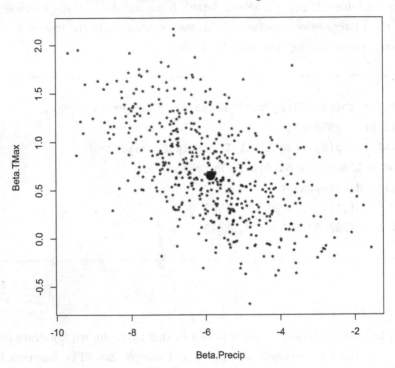

Figure 5.4: Scatter plot of slope parameters from the multivariate response L_1-regression fitting. The filled-in larger circle is the L_1-median.

We conclude this chapter with an example of how to compute an M_m-estimator and perform resampling for it. We consider the two-dimensional variable (`Precip`, `TMax`) as the response variable, and `TMin` as the explanatory variable along with an intercept term. This is a generalization of Example 2.7 presented in Chapter 2. We consider the case where the i-th observation variable $\mathbf{Y}_i \in \mathbb{R}^d$ with $d = 2$ in the present example. We assume that the L_1-median $\theta_i \in \mathbb{R}^d$ of $\mathbf{Y}_i$ may be expressed as a linear transformation of covariates $\mathbf{x}_i \in \mathbb{R}^p$, thus $\theta_i = B^T \mathbf{x}_i$ for some unknown parameter matrix $B \in \mathbb{R}^d \times \mathbb{R}^p$. This is a multivariate response version of *quantile regression* (Koenker and Bassett Jr, 1978). In our data example, we have $p = 2$, with

the first element of each covariate vector being 1 for the intercept term, and
the second element being the `TMin` observation. Thus, the parameter for
the present problem is the 2×2 matrix B. Note that the elements of the
response variable (`Precip`, `TMax`) are potentially physically related to each
other by the Clausius-Clapyron relation (see Dietz and Chatterjee (2014)),
and the data analysis of this section is motivated by the need to understand
the nature of this relationship conditional on the covariate `TMin`. We use the
function `L1Regression` in the package `UStatBookABSC` to obtain the estimator $\hat{B}$ minimizing

$$\Psi_n(B) = \sum_{i=1}^{n} |\mathbf{Y}_i - B^T \mathbf{x}_i|,$$

where recall that $|a| = [\sum a_k^2]^{1/2}$ is the Euclidean norm of a.

The following code implements the above task:

```
>DataY = cbind(CCU12_Precip$Precip, CCU12_Precip$TMax);
>DataX = cbind(rep(1, length(CCU12_Precip$Precip)),
                        CCU12_Precip$TMin)
>A2 = L1Regression(DataY, DataX)
>print(A2)
> print(A2)
$Iteration
[1] 4

$Convergence
[1] 0.0005873908

$BetaHat
            [,1]         [,2]
[1,]  164.803552  16.5318273
[2,]   -5.866605   0.6498325
```

We used default choices of the tuning parameters of `L1Regression`, which
can be queried using `?L1Regression`. In particular, the algorithm used inside
the function is iterative in nature, and if $\hat{B}_j$ is the estimate of B at iteration

step j, we compute the relative change in norm $|\hat{B}_{j+1} - \hat{B}_j|/|\hat{B}_j|$, where $|B|$ is the Euclidean norm of the vectorized version of B, ie, when the columns of B are stacked one below another to form a pd-length vector. We declare convergence of the algorithm if this relative change in norm is less than ϵ, and we use $\epsilon = 0.001$ here.

The above display shows that the function `L1regression` has a `list` as the output. The first item states that convergence occurred at the 4-th iteration step, and that the relative change in norm at the last step was 0.0005, and then the final $\hat{B}$ value is shown. Thus we have

$$\hat{B} = \begin{pmatrix} 164.803552 & 16.5318273 \\ -5.866605 & 0.6498325 \end{pmatrix}.$$

We also use the function `WLS` from the package `UStatBookABSC` to obtain a least squares estimator $\tilde{B}$ of B, for which the results are displayed below:

```
> A3 = WLS(DataY, DataX)
> print(A3);
          [,1]         [,2]
[1,] 145.412968 16.0526220
[2,]  -5.193363  0.6749197
```

Thus we have

$$\tilde{B} = \begin{pmatrix} 145.412968 & 16.0526220 \\ -5.193363 & 0.6749197 \end{pmatrix}.$$

It may be of interest to study the distribution of $\hat{B}[2,]$, the second row of $\hat{B}$, which contains the slope parameters for the covariate `TMin` for the two response variables. We use the generalized bootstrap with multinomial weights to get a consistent distributional approximation of $\hat{B}[2,]$. The following code implements this task:

```
>B = 500;
>Probabilities = rep(1, nrow(DataY))
```

```
>BootWeight = rmultinom(B, nrow(DataY), Probabilities);
>L1Regression.Boot = list();
>T2Effect = matrix( nrow = B, ncol = ncol(DataY));
>for (b in 1:B){
>Weights.Boot = diag(BootWeight[, b]);
>L1Regression.Boot[[b]] = L1Regression(DataY, DataX,
                              Weights = Weights.Boot);
>T2Effect[b, ] = L1Regression.Boot[[b]]$BetaHat[2,];
>}
```

In the above, we set the resampling Monte Carlo size at B = 500. The next two steps generates the multinomial weights for which we require the package MASS, which has been called inside UStatBookABSC anyway. The next steps implement a loop where L1Regression is repeatedly implemented with the resampling weights, the entire results stored in the list L1Regression.Boot and further, just the relevant slope parameters stored also in the matrix T2Effect.

We now present a graphical display of the above results with the code

```
> plot(T2Effect, xlab = "Beta.Precip", ylab = "Beta.TMax",
main = "Multivariate median regression", pch = "*")
>points(A2$BetaHat[2,1],A2$BetaHat[2,2], col = 2,
                 pch = 19, cex = 2);
>points(A3[2,1],A3[2,2], col = 3, pch = 22,
            cex = 2, bg = 3);
```

The above code is for plotting the resampling estimates of $\hat{B}[2,]$, stored in the matrix T2Effect. On this scatter plot, we overlay the original $\hat{B}[2,]$, as a big filled-in circle. The commands col =2 tells R to use red color for the point on the computer screen and in color print outs, pch = 19 tells it to depict the point with a filled-in red circle, and cex = 2 tells it to increase the size of the point. The output from the above code is given in Figure 5.4.

5.5 Exercises

1. Write a R function that can implement the Bayesian bootstrap on the L_1-regression example for the Kolkata precipitation data.

2. Write a function to implement the m-out-of-n bootstrap on the Oja-median for the Kolkata precipitation data. Discuss how can the results from this computation may be used to obtain a confidence region for the Oja-median.

3. Use resampling to obtain $1 - \alpha$ confidence intervals for each element of the L_1-regression slope matrix.

Bibliography

Abramovitch, L. and Singh, K. (1985). Edgeworth corrected pivotal statistics and the bootstrap. *The Annals of Statistics*, 13(1):116 – 132.

Arcones, M. A. (1996). The Bahadur-Kiefer representation for U - quantiles. *The Annals of Statistics*, 24(3):1400–1422.

Arcones, M. A., Chen, Z., and Gine, E. (1994). Estimators related to U-processes with applications to multivariate medians: Asymptotic normality. *The Annals of Statistics*, 22(3):1460 – 1477.

Arcones, M. A. and Gine, E. (1992). On the bootstrap of U- and V-statistics. *The Annals of Statistics*, 20:655 – 674.

Arcones, M. A. and Mason, D. M. (1997). A general approach to Bahadur-Kiefer representations for M-estimators. *Mathematical Methods of Statistics*, 6(3):267–292.

Athreya, K. B., Ghosh, M., Low, L. Y., and Sen, P. K. (1984). Laws of large numbers for bootstrapped U-statistics. *Journal of Statistical Planning and Inference*, 9(2):185 – 194.

Babu, G. J. and Singh, K. (1983). Inference on means using the bootstrap. *The Annals of Statistics*, 11(3):999 – 1003.

Babu, G. J. and Singh, K. (1984). On one term Edgeworth correction by Efron's bootstrap. *Sankhyā: Series A*, pages 219 – 232.

Babu, G. J. and Singh, K. (1985). Edgeworth expansions for sampling without replacement from finite populations. *Journal of Multivariate Analysis*, 17(3):261 – 278.

© Springer Nature Singapore Pte Ltd. 2018 and Hindustan Book Agency 2018
A. Bose and S. Chatterjee, *U-Statistics, Mm-Estimators and Resampling*, Texts and Readings in Mathematics 75, https://doi.org/10.1007/978-981-13-2248-8

Babu, G. J. and Singh, K. (1989a). A note on Edgeworth expansions for the lattice case. *Journal of Multivariate Analysis*, 30(1):27–33.

Babu, G. J. and Singh, K. (1989b). On Edgeworth expansions in the mixture cases. *The Annals of Statistics*, 17(1):443 – 447.

Bahadur, R. R. (1966). A note on quantiles in large samples. *The Annals of Mathematical Statistics*, 37(3):577 – 580.

Bentkus, V. and Götze, F. (1996). The Berry-Esseen bound for Student's statistic. *The Annals of Probability*, 24(1):491–503.

Bergsma, W. P. (2006). A new correlation coefficient, its orthogonal decomposition and associated tests of independence. *arXiv preprint math/0604627*.

Berk, R. H. (1966). Limiting behavior of posterior distributions when the model is incorrect. *The Annals of Mathematical Statistics*, 37(1):51 – 58.

Berk, R. H. (1970). Consistency a posteriori. *The Annals of Mathematical Statistics*, 41(3):894 – 906.

Berry, A. C. (1941). The accuracy of the Gaussian approximation to the sum of independent variates. *Transactions of the American Mathematical Socicty*, 49(1):122 – 136.

Bhattacharya, R. N. and Qumsiyeh, M. (1989). Second order and l^p-comparisons between the bootstrap and empirical Edgeworth expansion methodologies. *The Annals of Statistics*, 17(1):160 – 169.

Bhattacharya, R. N. and Rao, R. R. (1976). *Normal Approximation and Asymptotic Expansions*. Wiley Series in Probability and Mathematical Statistics. John Wiley, New York, USA.

Bickel, P. J. and Freedman, D. A. (1981). Some asymptotic theory for the bootstrap. *The Annals of Statistics*, 9:1196 – 1217.

Bickel, P. J. and Lehmann, E. L. (1979). Descriptive statistics for nonparametric models. IV. Spread. In Jurečková, J., editor, *Contributions to Statistics: Jaroslav Hájek Memorial Volume*, pages 33 – 40. Academia, Prague.

Bose, A. (1997). Bahadur representation and other asymptotic properties of M-estimates based on U-functionals. Technical report, Indian Statistical Institute, Kolkata, India.

Bose, A. (1998). Bahadur representation of M_m-estimates. *The Annals of Statistics*, 26(2):771 – 777.

Bose, A. and Babu, G. J. (1991). Accuracy of the bootstrap approximation. *Probability Theory and Related Fields*, 90(3):301 – 316.

Bose, A. and Chatterjee, S. (2001a). Generalised bootstrap in non-regular M-estimation problems. *Statistics and Probability Letters*, 55(3):319 – 328.

Bose, A. and Chatterjee, S. (2001b). Last passage times of minimum contrast estimators. *Journal of the Australian Mathematical Society-Series A*, 71(1):1 – 10.

Bose, A. and Chatterjee, S. (2003). Generalized bootstrap for estimators of minimizers of convex functions. *Journal of Statistical Planning and Inference*, 117(2):225 – 239.

Brillinger, D. R. (1962). A note on the rate of convergence of a mean. *Biometrika*, 49(3/4):574 – 576.

Castaing, C. and Valadier, M. (1977). *Convex Analysis and Measurable Multifunctions*, volume 580 of *Lecture Notes in Mathematics*. Springer-Verlag, Berlin.

Chatterjee, S. (1998). Another look at the jackknife: Further examples of generalized bootstrap. *Statistics and Probability Letters*, 40(4):307 – 319.

Chatterjee, S. (2016). *UStatBookABSC: A Companion Package to the Book "U-Statistics, M-Estimation and Resampling"*. R package version 1.0.0.

Chatterjee, S. and Bose, A. (2000). Variance estimation in high dimensional regression models. *Statistica Sinica*, 10(2):497 – 516.

Chatterjee, S. and Bose, A. (2005). Generalized bootstrap for estimating equations. *The Annals of Statistics*, 33(1):414 – 436.

Chaudhuri, P. (1992). Multivariate location estimation using extension of R-estimates through U-statistics type approach. *The Annals of Statistics*, 20(2):897 – 916.

Chaudhuri, P. (1996). On a geometric notion of quantiles for multivariate data. *Journal of the American Statistical Association*, 91(434):862 – 872.

Choudhury, J. and Serfling, R. J. (1988). Generalized order statistics, Bahadur representations, and sequential nonparametric fixed-width confidence intervals. *Journal of Statistical Planning and Inference*, 19(3):269
– 282.

Chow, Y. S. and Teicher, H. (1997). *Probability Theory: Independence, Inter-
changeability, Martingales*. Springer Texts in Mathematics. Springer, New
York, USA, 3rd edition.

Dalgaard, P. (2008). *Introductory Statistics with R*. Springer, 2nd edition.

Davison, A. C. and Hinkley, D. V. (1997). *Bootstrap Methods and their
Application*. Cambridge University Press, Cambridge, UK.

de la Peña, V. H. and Giné, E. (1999). *Decoupling: From Dependence to
Independence*. Springer, New York, USA.

Dehling, H., Denker, M., and Philipp, W. (1986). A bounded law of the
iterated logarithm for Hilbert space valued martingales and its application
to U-statistics. *Probability Theory and Related Fields*, 72(1):111 – 131.

Dietz, L. and Chatterjee, S. (2014). Logit-normal mixed model for Indian
monsoon precipitation. *Nonlinear Processes in Geophysics*, 21:934 – 953.

Durrett, R. (1991). *Probability: Theory and Examples*. Wadsworth, Pacific
Grove, CA, USA.

Dynkin, E. B. and Mandelbaum, A. (1983). Symmetric statistics, Poisson
point processes, and multiple Wiener integrals. *The Annals of Statistics*,
11(3):739 – 745.

Efron, B. (1979). Bootstrap methods: Another look at the jackknife. *The
Annals of Statistics*, 7(1):1 – 26.

Efron, B. (1982). *The Jackknife, the Bootstrap and Other Resampling Plans*.
SIAM, Philadelphia, USA.

Efron, B. and Tibshirani, R. J. (1993). *An Introduction to the Bootstrap*.
Chapman & Hall/CRC press, Boca Raton, USA.

Esseen, C.-G. (1942). On the Liapounoff limit of error in the theory of probability. *Arkiv för Matematik, Astronomi och Fysik*, 28A(2):1 – 19.

Esseen, C.-G. (1945). Fourier analysis of distribution functions. A mathematical study of the Laplace-Gaussian law. *Acta Mathematica*, 77(1):1 – 125.

Esseen, C.-G. (1956). A moment inequality with an application to the central limit theorem. *Scandinavian Actuarial Journal (Skandinavisk Aktuarietidskrift)*, 39(3-4):160 – 170.

Falk, M. (1992). Bootstrapping the sample quantile: A survey. In Jöckel, K.-H., Rothe, G., and Sendler, W., editors, *Bootstrapping and Related Techniques: Proceedings of an International Conference, Held in Trier, FRG, June 4–8, 1990*, pages 165 – 172. Springer, Berlin, Heidelberg.

Falk, M. and Reiss, R. D. (1989). Weak convergence of smoothed and nonsmoothed bootstrap quantile estimates. *The Annals of Probability*, 17(1):362 – 371.

Freedman, D. A. (1981). Bootstrapping regression models. *The Annals of Statistics*, 9(6):1218 – 1228.

Ghosh, M., Parr, W. C., Singh, K., and Babu, G. J. (1984). A note on bootstrapping the sample median. *The Annals of Statistics*, 12(3):1130–1135.

Grams, W. F. and Serfling, R. J. (1973). Convergence rates for U-statistics and related statistics. *The Annals of Statistics*, 1(1):153 – 160.

Gregory, G. G. (1977). Large sample theory for U-statistics and tests of fit. *The Annals of Statistics*, 5(1):110–123.

Haberman, S. J. (1989). Concavity and estimation. *The Annals of Statistics*, 17(4):1631 – 1661.

Hájek, J. (1961). Some extensions of the Wald - Wolfowitz - Noether Theorem. *The Annals of Mathematical Statistics*, 32(2):506 – 523.

Hall, P. (1986). On the bootstrap and confidence intervals. *The Annals of Statistics*, 14(4):1431–1452.

Hall, P. (1988). Theoretical comparison of bootstrap confidence intervals. *The Annals of Statistics*, 16(3):927 – 953.

Hall, P. (1992). *The Bootstrap and Edgeworth Expansion*. Springer-Verlag, New York, USA.

Hall, P. (2003). A short prehistory of the bootstrap. *Statistical Science*, 18(2):158 – 167.

Hall, P. and Martin, M. A. (1991). On the error incurred using the bootstrap variance estimate when constructing confidence intervals for quantiles. *Journal of Multivariate Analysis*, 38(1):70 – 81.

Heiler, S. and Willers, R. (1988). Asymptotic normality of R-estimates in the linear model. *Statistics*, 19(2):173 – 184.

Helmers, R. (1991). On the Edgeworth expansion and the bootstrap approximation for a Studentized U-statistic. *The Annals of Statistics*, 19:470 – 484.

Helmers, R. and Hušková, M. (1994). Bootstrapping multivariate U-quantiles and related statistics. *Journal of Multivariate Analysis*, 49(1):97–109.

Hjort, N. L. and Pollard, D. (1993). Asymptotics for minimizers of convex processes. Preprint series. Statistical Research Report, Matematisk Institutt, Universitet i Oslo.

Hodges, J. L. and Lehmann, E. L. (1963). Estimates of location based on rank tests. *The Annals of Mathematical Statistics*, 34(2):598 – 611.

Hoeffding, W. (1948). A class of statistics with asymptotically normal distribution. *The Annals of Mathematical Statistics*, 19:293 – 325.

Hoeffding, W. (1961). The strong law of large numbers for U-statistics. Institute of Statistics mimeo series 302, University of North Carolina, Chapel Hill, USA.

Hollander, M. and Wolfe, D. A. (1973). *Nonparametric Statistical Methods*. John Wiley and Sons, New York, USA.

Hubback, J. A. (1946). Sampling for rice yield in Bihar and Orissa. *Sankhyā*, pages 281 – 294. First published in 1927 as Bulletin 166, Imperial Agricultural Research Institute, Pusa, India.

Huber, P. J. (1964). Robust estimation of a location parameter. *The Annals of Mathematical Statistics*, 35(1):73 – 101.

Huber, P. J. (1967). The behavior of maximum likelihood estimates under nonstandard conditions. In *Proceedings of the Fifth Berkeley Symposium on Mathematical Statistics and Probability*, volume 1, pages 221 – 233.

Hušková, M. and Janssen, P. (1993a). Consistency of the generalized bootstrap for degenerate U-statistics. *The Annals of Statistics*, 21(4):1811 – 1823.

Hušková, M. and Janssen, P. (1993b). Generalized bootstrap for Studentized U-statistics: A rank statistic approach. *Statistics and Probability Letters*, 16(3):225–233.

Jurečková, J. (1977). Asymptotic relations of M-estimates and R-estimates in linear regression model. *The Annals of Statistics*, 5(3):464 – 472.

Jurečková, J. (1983). Asymptotic behavior of M-estimators of location in nonregular cases. *Statistics & Risk Modeling*, 1(4 - 5):323 – 340.

Jurecková, J. and Sen, P. K. (1996). *Robust Statistical Procedures: Asymptotics and Interrelations*. John Wiley & Sons.

Kemperman, J. H. B. (1987). The median of a finite measure on a Banach space. In Dodge, Y., editor, *Statistical Data Analysis Based on the L_1-Norm and Related Methods (Neuchâtel, 1987)*, pages 217–230. North Holland, Amsterdam, Holland.

Kiefer, J. (1967). On Bahadur's representation of sample quantiles. *The Annals of Mathematical Statistics*, 38(5):1323 – 1342.

Knopp, K. (1923). *Theory and Application of Infinite Series*. Blackie and Sons Ltd., London, UK.

Koenker, R. and Bassett Jr, G. (1978). Regression quantiles. *Econometrica*, pages 33 – 50.

Kolassa, J. E. and McCullagh, P. (1990). Edgeworth series for lattice distributions. *The Annals of Statistics*, 18(2):981 – 985.

Korolev, V. Y. and Shevtsova, I. G. (2010). On the upper bound for the absolute constant in the Berry-Esseen inequality. *Theory of Probability and Its Applications*, 54(4):638 – 658.

Korolyuk, V. S. and Borovskich, Y. V. (1993). *Theory of U-Statistics.* Springer, New York, USA.

Lee, A. J. (1990). *U-Statistics: Theory and Practice.* Marcel Dekker Inc., New York.

León, C. A. and Massé, J.-C. (1993). La médiane simpliciale d'Oja: Existence, unicité et stabilité. *The Canadian Journal of Statistics/La Revue Canadienne de Statistique,* 21(4):397 – 408.

Liu, R. Y. (1990). On a notion of data depth based on random simplices. *The Annals of Statistics,* 18(1):405 – 414.

Lyapunov, A. (1900). Sur une proposition de la théorie des probabilités. *Bulletin de l'Académie Impériale des Sciences,* 13(4):359 – 386.

Lyapunov, A. (1901). Nouvelle forme du théoreme sur la limite de probabilité. *Memoires de l'Academe de St-Pétersbourg,* Volume 12.

Mahalanobis, P. C. (1940). A sample survey of the acreage under jute in Bengal. *Sankhyā,* 4(4):511 – 530.

Mahalanobis, P. C. (1944). On large-scale sample surveys. *Philosophical Transactions of the Royal Society of London B: Biological Sciences,* 231(584):329 – 451.

Mahalanobis, P. C. (1945). Report on the Bihar crop survey: Rabi season 1943-44. *Sankhyā,* 7(1):29 – 106.

Mahalanobis, P. C. (1946a). Recent experiments in statistical sampling in the Indian Statistical Institute. *Journal of the Royal Statistical Society: Series A (Statistics in Society),* 109(4):325 – 378. Reprinted, including discussion, in Sankhyā, volume 20, (1958) 329–397.

Mahalanobis, P. C. (1946b). Sample surveys of crop yields in India. *Sankhyā,* pages 269 – 280.

Mallows, C. L. (1972). A note on asymptotic joint normality. *The Annals of Mathematical Statistics,* 43(2):508 – 515.

Mammen, E. (1993). Bootstrap and wild bootstrap for high dimensional linear models. *The Annals of Statistics,* 21(1):255 – 285.

Maritz, J. S., Wu, M., and Stuadte, R. G. (1977). A location estimator based on a U-statistic. *The Annals of Statistics*, 5(4):779 – 786.

Massart, P. (1990). The tight constant in the Dvoretzky-Kiefer-Wolfowitz inequality. *The Annals of Probability*, 18(3):1269–1283.

Matloff, N. (2011). *The Art of R Programming: A Tour of Statistical Software Design*. No Starch Press, San Francisco, USA.

Miller, R. G. (1964). A trustworthy jackknife. *The Annals of Mathematical Statistics*, 35(4):1594–1605.

Niemiro, W. (1992). Asymptotics for M-estimators defined by convex minimization. *The Annals of Statistics*, 20(3):1514–1533.

Oja, H. (1983). Descriptive statistics for multivariate distributions. *Statistics and Probability Letters*, 1(6):327 – 332.

Oja, H. (1984). Asymptotical properties of estimators based on U-statistics. preprint, Dept. of Applied Mathematics, University of Oulu, Finland.

Petrov, V. V. (1975). *Sums of Independent Random Variables*. Springer, New York, USA.

Pollard, D. (1991). Asymptotics for least absolute deviation regression estimators. *Econometric Theory*, 7(02):186 – 199.

Præstgaard, J. and Wellner, J. A. (1993). Exchangeably weighted bootstraps of the general empirical process. *The Annals of Probability*, 21(4):2053–2086.

Quenouille, M. H. (1949). Approximate tests of correlation in time-series. *Journal of the Royal Statistical Society: Series B (Methodological)*, 11(1):68–84.

Rockafellar, R. T. (1970). *Convex Analysis*. Princeton Mathematical Series. Princeton University Press.

Rousseeuw, P. J. (1985). Multivariate estimation with high breakdown point. In Grossman, W., Pflug, G., Vincze, I., and Wertz, W., editors, *Mathematical Statistics and Applications*, pages 283 – 297. Reidel Publishing Company, Dordrecht.

Serfling, R. J. (1980). *Approximation Theorems of Mathematical Statistics.* John Wiley & Sons, New York, USA.

Shao, J. and Tu, D. (1995). *The Jackknife and Bootstrap.* Springer-Verlag, New York, USA.

Sherman, R. P. (1994). Maximal inequalities for degenerate U-processes with applications to optimization estimators. *The Annals of Statistics*, 22(1):439 – 459.

Shevtsova, I. (2014). On the absolute constants in the Berry-Esseen type inequalities for identically distributed summands. *Doklady Mathematics*, 89(3):378 – 381.

Shorack, G. R. (2000). *Probability for Statisticians.* Springer Texts in Statistics. Springer, New York, USA.

Singh, K. (1981). On the asymptotic accuracy of Efron's bootstrap. *The Annals of Statistics*, 9(6):1187 – 1195.

Small, C. G. (1990). A survey of multidimensional medians. *International Statistical Review*, 58(3):263 – 277.

Smirnov, N. V. (1949). Limit distributions for the terms of a variational series. *Trudy Matematicheskogo Instituta im. VA Steklova*, 25:3 – 60.

Smirnov, N. V. (1952). Limit distributions for the terms of a variational series. *American Mathematical Society Translation Series*, 11(1):82 – 143.

Student (1908). The probable error of a mean. *Biometrika*, 6(1):1–25.

Tukey, J. W. (1958). Bias and confidence in not quite large samples (abstract). *The Annals of Mathematical Statistics*, 29(2):614–614.

Tukey, J. W. (1975). Mathematics and the picturing of data. In *Proceedings of the International Congress of Mathematicians*, volume 2, pages 523 – 531.

van der Vaart, A. W. and Wellner, J. A. (1996). *Weak Convergence and Empirical Processes: With Applications to Statistics.* Springer, New York, USA.

Wagner, T. J. (1969). On the rate of convergence for the law of large numbers. *The Annals of Mathematical Statistics*, 40(6):2195 – 2197.

Wasserman, L. (2006). *All of Nonparametric Statistics*. Springer, New York, USA.

Wu, C. F. J. (1990). On the asymptotic properties of the jackknife histogram. *The Annals of Statistics*, 18(3):1438 – 1452.

Author Index

© Springer Nature Singapore Pte Ltd. 2018 and Hindustan Book Agency 2018
A. Bose and S. Chatterjee, *U-Statistics, Mm-Estimators and Resampling*, Texts
and Readings in Mathematics 75, https://doi.org/10.1007/978-981-13-2248-8

Subject Index

© Springer Nature Singapore Pte Ltd. 2018 and Hindustan Book Agency 2018
A. Bose and S. Chatterjee, *U-Statistics, Mm-Estimators and Resampling*, Texts and Readings in Mathematics 75, https://doi.org/10.1007/978-981-13-2248-8

Texts and Readings in Mathematics

Printed in the United States
By Bookmasters